Just some of the wonders of Rioja

1 The historic bodegas of the Barrio de la Estación in Haro (see page 20)

2 The Logroño tapas trail (see page 90)

3 Santiago Calatrava's Ysios (see page 120)

4 Marqués de Riscal (see page 115)

5 The beautiful walled town of Labastida (see page 93)

6 The cooperative Fincas de Azabache – another side of Rioja (see page 105)

7 Vivanco Museum of Wine Culture (see page 90)

8 The ancient monasteries at San Millán de la Cogolla (see page 102 and picture above)

All prices are correct at time of going to
press, but are subject to change.

Published 2024 by Académie du Vin Library Ltd
academieduvinlibrary.com
Founders: Steven Spurrier and Simon McMurtrie

Publisher: Hermione Ireland
Series editor: Adam Lechmere
Editor: Jon Richards
Copy editor: David Tombesi-Walton
Design: Dan Prescott and Gary Hyde, Imago Create
Maps supplied by Cosmographics
Index: Marie Lorimer
Proofreader: Jenny Sykes
ISBN: 978-1-917084-50-5
Printed and bound in the EU
© 2024 Académie du Vin Library Ltd

Rioja

Fintan Kerr

THE SMART TRAVELLER'S
WINE GUIDE

Contents

Foreword   8

Map of Rioja   10
Introduction   12

History   14

Rioja classifications and how they work   26
Old vines and why they matter   32
Rioja's greatest wines   36

Geography   46

Wine styles   58
Grape varieties   68
How to read a Rioja label   76
Great Rioja vintages   80

Visiting Rioja   84
The towns and villages of Rioja   92
Wine routes   100

The Guide 108

The best bodegas for tours and tastings 110
The best wine hotels 122
Fine dining in Rioja 134
The best bodega restaurants 140
The best tapas in Logroño 149
Where to eat and drink in Haro 155
The best wine shops in Rioja 162
Wines to look out for 166

Glossary 174
Further reading 176
Index 178
Acknowledgements 182

Ancient and modern Rioja: the titanium exuberance of hotel Marqués de Riscal a step away from the 16th-century church of San Andrés, with the Sierra de Cantabria looming behind

Gnarled old vines at Viñedos del Contino, Laguardia

Foreword

Every great wine region has its famous domaines and vineyards. There's no doubt that Rioja's finest wines are world class, but international recognition, like our nation's football success, has been relatively recent. For the visitor, this is a very good thing.

Wine has been made in Rioja for a very long time. There is some debate over whether the spirit of the 'true' Rioja was captured in the 18th century, and lost since – there are no wines around from that era, so we'll never know. But we all accept that modern Rioja was born in the 19th century. Its centre was Haro, home to some of the most historic wineries not only in Rioja but in all of Spain.

Some of them (like CVNE, which was founded in 1879 by my ancestors, the Real de Asúa brothers) are still owned by their founding families. These historic bodegas, from Haro to Elciego and beyond, are beautiful, unique and, importantly, open to visitors.

Equally interesting are the new domaines, more often than not created by vignerons that above all seek to interpret their vineyards as faithfully as possible. The best produce the most extraordinary wines.

In his 1990 classic, Adventures on the Wine Route, the great American importer Kermit Lynch recommended visiting Beaujolais and Alsace before going to, say, the Côte d'Or or the Médoc. He said the wines would still be outstanding, but more affordable; the locals would be welcoming, and the visiting wine enthusiast would not find herself crowded out by the masses or turned away by jaded winery employees.

He forgot to mention Rioja, which has what we all seem to crave nowadays: authenticity. As you walk through villages that smell of wine over harvest time, you will meet vignerons and winemakers, charming bar and restaurant staff, all the time reassured that Rioja is not a tourist destination but a thriving community centred on wine.

This book captures the essence of what makes Rioja a place that every wine lover should visit at least once. It should be in every wine lover's suitcase.

Victor Urrutia
Chief Executive
CVNE
July 2024

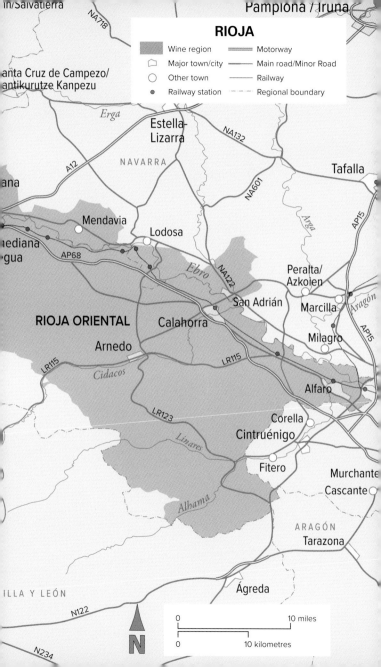

Introduction

Rioja is as familiar to the wine lover as Islay to the whisky aficionado. From the humblest crianza to the most exalted single-vineyard cuvée, this great wine region excels at every level.

This is one of the most beguiling of all Spanish regions, a place of medieval hilltop towns with cobbled streets winding between houses that seem not to have changed in centuries; of boisterous festivals whose origins are lost in time, of religious rites that go back to the crusades.

None of this is laid on for tourists. These are working towns – those cobbled streets ring to the sound of people heading to the office or the winery, not ambling groups fresh off a cruise ship. If you visit at harvest time, you're likely to be held up behind farmers sitting on their tractors with a whole year's work on the trailer.

Part of the joy of Rioja is its mix of artisanry and international sophistication. Here, Renaissance churches stand shoulder to shoulder with modernist masterpieces, traditional wood-and-stone inns serve pigs' trotters or *Patorrillo a la Riojana* next door to Michelin-starred restaurants like Haro's Nublo. And don't forget the tapas in Logroño.

Riojanos have a streak of radicalism. Look at the avant-garde architecture (Gehry's Marqués de Riscal or Calatrava's Ysios are just two among many extraordinary buildings), the array of concrete tanks at Lalomba (see pages 24–25), or how appellation rules that seemed set in stone have been comprehensively updated. All this was created by people with their eyes firmly on the future.

This book is aimed at anyone with an urge to find out more about this legendary corner of the wine world. We'll tell you which wineries to visit and when, the best restaurants, where to find the choicest tapas, the best wine shops – and we'll tell you how it all came to be. Above all, if you love the smell of a barrel cellar, this is the guide for you.

History

Rioja in ancient times

As with many of the world's wine regions, it is possible that we'll never truly know how winemaking first came to be practised in Rioja. The Phoenicians are believed to have introduced viticulture and grapevines in other parts of Spain, such as Catalonia and Andalusia, but they never ventured this far north. Instead, various Celtic tribes inhabited the rolling hills of Rioja – until the Romans arrived and conquered the region, uncharacteristically late, at the beginning of the 2nd century BCE.

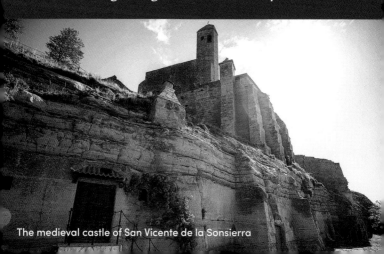

The medieval castle of San Vicente de la Sonsierra

Ancient ruins near Logroño give us an indication of the impact that the Romans had on winemaking here. Vast quantities of wine were produced in the area, much of it for the legions and the growing local population. Wine was of vital importance to the Roman army – both for morale and for its role in helping to sanitize drinking water. Vineyards were planted, mostly from the same grape varieties that are still grown in Rioja today, and the first true wineries sprang up around them. The Romans also brought with them the technology of the age: fermentations were generally completed in open stone troughs known as *lagares*, before being transferred to clay amphorae for storage and transportation. Even today, it is possible to find wine produced this way in rural parts of Rioja, where it is known as *vino de cosechero*, or 'grower's wine', and many vineyards around Rioja still have ancient stone *lagares*, standing as eternal reminders of the origins of local wine production.

Empires, however, rise and fall; the Roman Empire collapsed towards the end of the 4th century CE, leaving Spain vulnerable, lawless and ripe for the taking. First came the Visigoths, a Germanic people who ruled Rioja and much of Spain until the Moorish invasion of 711 CE. By the time the Moors reached Rioja in the far north of the country, the wine industry was a key part of the local economy. The strict rules imposed by the Moors on the consumption of alcohol meant that progression in winemaking slowed to a halt. Distillation continued to be widely practised for the production of perfumes and cosmetics, but wine production was limited to individual farmers and not available on a commercial scale. Fortunately, this period was short-lived: the long wars against the Moors and the reclamation of Spanish territories began in nearby Asturias, and by 923 CE, Rioja had been reclaimed by Sancho I, the king of Pamplona.

'Many vineyards still have ancient stone *lagares*'

Medieval Rioja

During the Middle Ages, Rioja increased its production of wine, if not its international reputation. It took until the 15th century to finally drive the Moors completely out of Spain, but by that point, Rioja – or the Province of Logroño, as it was then known – had been subject to almost five centuries of comparatively stable rule.

Wine was an essential part of a daily diet in Rioja. Importantly, its consumption was not limited to the noble and landowning classes, but it extended to workers in the fields and labourers, for whom it was an important source of calories and sustenance. Vines were planted across the entire region, and production levels soared.

Riojan wine at the time would have been fairly simple, mostly produced in stone *lagares* using methods introduced by the Romans. *Lagares* were usually carved out of the limestone rock found on the banks of the River Ebro, close to the vineyards, with stone bases used for stepping on to slowly crush the grapes. Narrow channels were carved at a slight inclination, allowing the freshly squeezed juices

A guardaviñas. Fewer than 100 of these typical c19th vineyard shelters remain

to run down into a deep tank in a cave below, known at the time as a *torco*, where alcoholic fermentation would take place. These are still visible in Sonsierra today, as are the distinctive conical shelters, built to protect vineyard workers from the rain or excessive heat.

As productive as Rioja was during this period, its wines were made almost entirely for local consumption. Spanish law was protectionist, especially for wine, with edicts stating that only local wine could be consumed within each region. Travellers – notably thirsty pilgrims on the great Camino de Santiago – did far more for the international reputation of Rioja at the time than its limited exports.

The rebirth of Rioja

This pattern continued throughout the region for hundreds of years. Development was slow, and it wasn't until the second half of the 19th century that Rioja began to be recognized as a wine region of international importance.

In terms of agricultural production and methods, Rioja had looked north for inspiration for some time. Bilbao, an important port, was the entry point to Spain for products such as fruit, molasses, sugars and spices, as well as wines from other regions, notably the Canary Islands and Bordeaux. A good deal of Bordeaux wine was sold through Spain and imported into Britain at a lower rate of tax (under the guise of being Spanish), before being labelled as claret and sold on. Both economically and geographically, Bordeaux and Rioja have always been close: for hundreds of years, Rioja was easier to get to than Paris.

In 1785, Manuel Quintano, a priest of Riojan descent, went to Bordeaux to study winemaking; two years later, he published a treatise recommending the adoption of certain Bordeaux techniques. It took more than a decade for these to be approved by the Rioja authorities, but the Napoleonic Wars (1803–15) and political upheavals in Spain meant that Quintano's ideas were never properly put into practice.

Almost 50 years later, Eugenio de Garagarza – the

director of Álava's Model Farm, which focused on the development of agricultural engineering – led a second attempt to learn how to make wine following Bordelais methodology. In 1860, Garagarza, himself a graduate of a French agricultural school, recommended hiring a winemaker from Bordeaux to help teach 'the secrets of the Médoc' to producers in Rioja. Under this scheme, Jean Pineau, cellar master at Château Lanessan, was employed by Marqués de Riscal and spent seven years teaching Bordeaux methods to local winemakers.

The official scheme, which was financed by the regional government, fell victim to disputes over cost, but the French connection was strong. The Marqués de Riscal himself, Camilo Hurtado de Amézaga, a diplomat and free thinker, had lived in Bordeaux in self-imposed exile since the 1830s, and his knowledge of the region and its wines led him to continue the project under his own private banner. He re-hired Pineau, built a winery to the cellar master's specifications and gave him free rein. Within a decade, the wines of Marqués de Riscal were gaining international attention.

As with all new ventures, these winemaking techniques were scoffed at initially; however, within a few decades, every winery in Rioja was following the same methods: fermentation in large oak vats, followed by maturation in smaller oak barrels, or the *méthode bordelaise*. The major difference between Bordeaux and Rioja was the latter's reliance on oak from the Americas – Spain's colonial links meant strong trading routes already existed across the Atlantic, and American oak was considerably cheaper than French wood. The preponderance of American oak in Rioja is arguably more the result of economic necessity than any natural affinity of Tempranillo with the sweeter, more fragrant American wood.

The Bordeaux connection was to be cemented by the advent of phylloxera. The discovery of the insect in French vineyards in the 1860s was to have far-reaching effects for Rioja – on both the prosperity of the region and the style of its wines. Phylloxera, introduced to Europe via American vine cuttings, is a louse that lives on a vine's roots, eating into them and gradually killing the vine. In the second half of the 19th century, the entirety of the French vineyard – except for a few pockets of sandy soils in which the louse can't live – was destroyed.

As the vineyards in Bordeaux withered, many of their winemakers and vignerons made the journey across the Pyrenees, settling in Rioja and bringing their considerable knowledge and expertise with them. The local cities of San Sebastián and Bilbao were suddenly awash with excellent wine; France, with its own production still devastated, welcomed this new source of exported wine.

In the last half of the 19th century, the ties between Bordeaux and Rioja became ever stronger. Millions of litres of Tempranillo were exported in bulk to Bordeaux and sold on as claret. Trade agreements meant lower duties for Spanish imports to France, and Bordeaux merchants

Oak fermentation vats at Marqués de Riscal

appointed agents throughout Rioja, bought vineyards and produced wine using their own techniques and to their preferred style. In 1863, a railway station was built at Haro to service this booming trade. Clustered around this unassuming terminus are some of the most eminent names in Rioja wine: CVNE, Bodegas Bilbaínas, Gómez Cruzado, La Rioja Alta, Muga, Bodegas Roda and López de Heredia. Every year, various tastings are held to honour the winemaking history of the Barrio de la Estación.

Phylloxera eventually made its way across the border to Spain, but it didn't arrive in Rioja until the 1890s, by which time a solution had been found: vines could be grafted on to American roots, because American vines are resistant to the insect. Across Europe, vineyards were slowly replaced. Rioja had already established itself internationally, and the station at Haro, with direct routes to major Atlantic ports, allowed it to build on and expand its reputation. The city of Haro was officially inaugurated in 1891, cementing its importance as a hub for the rapidly expanding region. Winemaking expertise and ambition spread throughout Rioja.

Rioja in the 20th century

The first ten years of the 20th century were spent replanting vineyards decimated by phylloxera. Then World War I destroyed stability in Europe and obliterated export markets. Further upheavals followed over the next three or four decades: international economic depression, the Spanish Civil War and World War II. Vineyards were destroyed by war, or replanted with grain to ease the crisis in food production, and it wasn't until the 1960s and 1970s that Rioja began to regain some sort of equilibrium. In this it was able to rely on its international reputation – and, crucially, on the fact that in 1926 it had become the first Spanish region to be demarcated as a quality producer of wine. In that year, the Consejo Regulador (Regulating Council) was set up to oversee production and control the

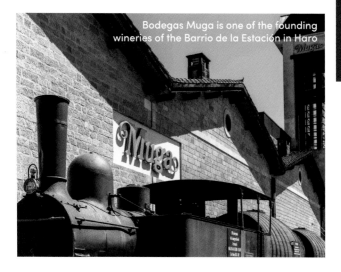

Bodegas Muga is one of the founding wineries of the Barrio de la Estación in Haro

use of the Rioja name on the region's wines. More than half a century later, in 1991, Rioja became the first DOCa – *denominación de origen calificada*.

Some of the most famous Rioja vintages took place during the 1950s and 1960s, particularly 1952, 1955 and 1964. The combination of rebound growth after World War II and the increasing support from international markets, particularly Britain, allowed Rioja to slowly grow in fame and fortune; some legendary bottles from top producers such as López de Heredia and La Rioja Alta only added to its renown. However, as is so often the case, this era of popularity was followed by an unfortunate period of overconfidence and overproduction. High yields were favoured to keep up with increasing demand, but this reduced the quality of the fruit, and many of the wines lacked the fruit concentration to manage the traditional long ageing processes. The emergence of DO Ribera del Duero during this period, particularly during the 1980s, led to a short-term loss in reputation for Rioja.

Rioja in the 21st century

Rioja has undergone significant changes since the turn of the century. Along with other wine regions, it has had to navigate a trend for overextracted, heavily oaked wines made to appeal to an international audience in thrall to a style largely popularized by the influential American critic Robert Parker. It is currently finding a balance between a more modern, elegant approach and the traditions and history for which it is famous. Rioja now makes wines in a variety of styles – from classic oaked reds, to lean, textured, lighter expressions, often produced with minimal oak ageing. In the hands of the best winemakers, the character of Tempranillo – and indeed of the many international grapes that are now being experimented with – and of the Rioja terroir will shine through.

The barrel cellar at Bodegas Vivanco

Architect Iñaki Aspiazu's dramatic concrete walkways at Bodegas Baigorri in Rioja Alavesa

Like a modernist cathedral: some of the 48 cement tanks at Ramón Bilbao's Lalomba, where single vineyard wines see little or no oak

Rioja classifications and how they work

As always in Spain, politics is never far from the surface. As recently as 2015, Rioja was riven with disagreement; the crux of the argument could be summed up as, 'Single vineyard good; blending bad.' This is a gross simplification of a very complicated situation, but it has a core of truth.

Vineyards in Sajazarra, in the northwest of Rioja Alta

There was impassioned debate: many believed that Rioja's rules should be updated. The traditional idea that quality was based on the length of time a wine spent in barrel, with little emphasis placed on the primacy of plot, was out of step with the rest of the wine world. The conversation found its focus in the 'Matador Manifesto', a 2015 document signed by 150 producers calling for Rioja's terroirs to be celebrated by a change in the rules that would allow single-vineyard and village names to be included on labels.

Some felt that this cast aside an honourable tradition. The eminent critic Pedro Ballesteros Torres MW said it was sad that Rioja was being criticized 'for its classification based on minimum ageing times'. He pointed out that the modern trend to see blending as somehow a negation of terroir was illogical: 'You will never get a good blend if you don't know your terroir intimately.'

Long speeches were made on either side of the argument, and finally the Consejo Regulador agreed there would be changes. The result is a new pyramid of quality in Rioja that exists alongside the old quality levels of **crianza**, **reserva** and **gran reserva**. This new classification is focused on terroir, and it is constantly shifting. (The most recent adjustment at time of writing was in February 2024.)

There are now three geographical classifications. **Vino de zona** allows producers to label their wines Rioja Alavesa, Rioja Alta or Rioja Oriental. One level up is village wine – **vino de pueblo** (formerly known as vino de municipio) – which allows producers to put one of 144 villages on their label. Then there is **viñedo singular**, by which producers can reference a single plot, where the vines must be at least 35 years old and harvested by hand.

Wine bureaucracy is always complicated – and never more so than in Spain. There has been controversy over the vino de pueblo classification because it requires both vineyard and winery to be in the same village; consequently, there will soon be a new designation on labels: **viñedo en...** ('vineyard in...'), meaning that the wine comes from a single vineyard but is vinified somewhere else.

The discussion continues, of course, but things are much calmer now. Those venerable bodegas, past masters of the great art of blending, are for the most part satisfied that they are not being dismissed as inferior; and those who make wine from single vineyards and villages are happy that they can proclaim it on their label. Take-up of the new regulations has been respectable: by July 2023, there were 148 registered viñedo singular wines, produced by 90 different wineries. That number will only grow.

Barrel-ageing classifications

Under regulations first set out in 1926, there are three basic levels for barrel ageing. Note that these are minimum ageing times; many bodegas will keep their finest wines several years longer in barrel and bottle.

• **Crianza** wines must spend a minimum of one year in oak barrel and one year in bottle before release.
• **Reserva** wines must spend one year in oak barrel and two in bottle before release.
• **Gran reserva** wines must spend at least two years in oak barrel and three years in bottle before release.

'Wine bureaucracy is always complicated – and never more so than in Spain'

Geographical classifications

• The **vino de zona** classification allows producers to label their wines Rioja Alavesa, Rioja Alta or Rioja Oriental.
• **Vino de pueblo** (recently changed from vino de municipio) allows producers to put one of 144 villages on their label.
• Under the **viñedo singular** banner, producers can reference a single vineyard or plot. The vines must be at least 35 years old and harvested by hand.
• **Viñedo en...** is similar to vino de pueblo, but the wine does not have to be vinified in the same village.

Thousands of bottles maturing in
the cellars at Contino, Laguardia

The Consejo Regulador – How Rioja is governed

The *Consejo Regulador de la Denominación de Origen Calificada Rioja (Rioja DOCa Control Board)* is the guiding force behind Spain's most renowned wine region: Rioja. Established in 1925 to champion excellence, the Rioja DOCa Control Board meticulously oversees every step of the winemaking journey from vine to bottle to ensure the utmost quality.

The official Rioja back label is a guaranteed mark of quality, indicating that the wine has been produced, aged and bottled according to the strict] set forth by the Rioja DOCa Control Board. This process includes evaluating grape varieties, ageing methods and labelling practices in order to ensure consistency and authenticity across all Rioja wines. This rigorous process reflects the region's esteemed winemaking traditions and reputation for excellence.

Dedicated to innovation, the Rioja DOCa Control Board fuels advancements in the science and study of wine and winemaking, fostering sustainable practices and enhancing wine quality. By sharing cutting-edge research and techniques, it empowers local winemakers to push boundaries and refine their craft. This commitment to innovation ensures that Rioja wines continue to evolve and maintain their competitive edge in the wine industry.

The Rioja DOCa Control Board also works to promote Rioja wines

around the world and increase knowledge and awareness of wines from the region. From large-scale trade shows to captivating events, its global marketing initiatives celebrate Rioja's rich heritage and elevate its prominence in the wine community.

Through its own dedicated online platform *Rioja Wine Academy,* the Rioja DOCa Control Board aims to make wine education accessible and enjoyable for everyone. For example, the 'Rioja Enthusiast' course – one of many available – is a free introductory course for wine lovers who want to learn more about Rioja and its wines, with no prior knowledge of wine required. In just four hours, the course covers the fundamentals of Rioja wine, including topics like reading labels and pairing wines. Successful completion of the course will earn students an official certification. Learn more about Rioja at: riojawineacademy.com

Through these multifaceted efforts, the Rioja DOCa Control Board upholds the high standards and global reputation of Rioja wines, ensuring they continue to be recognized as some of the finest in the world.

Content supported by the Consejo Regulador

Old vines and why they matter

Looking after old vines has always been important, but with the increased focus on environmental awareness and sustainability, the preservation of ancient vineyards has assumed an even greater urgency for the global winemaking community.

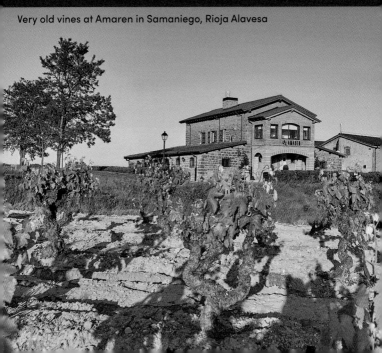

Very old vines at Amaren in Samaniego, Rioja Alavesa

One of the difficulties with old vines is that the definition of 'old' differs from country to country and region to region. According to The Old Vine Conference, which keeps a global register of old vines, 35 years is generally considered the minimum age for an old vine. Phylloxera devastated vineyards in the late 19th and early 20th centuries, but there are still pockets of old vines around the world – often in the sandy soils that are resistant to the disease. You can find them in California, Australia and South Africa; and there are some ancient vineyards in parts of southwest France (but few in Bordeaux) and Italy. Wherever wine is made, you will find vineyards of gnarly, hollowed-out old vines that are still capable of producing grapes.

Some 20% of Rioja's vineyards were planted before 1986. The importance of these old vines was recognized in Rioja's new classifications (see page 26), which include viñedo singular (single vineyard); the inclusion of this term on a label means the wine is made from vines that are at least 35 years old. Look out, too, for labels carrying the term 'viñas viejas' or 'cepas viejas', or even a year when the vines were planted (*año de plantación*).

There are many vines much older than 35 years in parts of Rioja. Some vineyards in Rioja Oriental escaped the phylloxera crisis, and there are some very old Garnacha bush vines there, such as Bodegas Palacios Remondo's 90-year-old Garnacha in Alfaro, which goes into its Viñas Viejas de la Propiedad. There are also plots of old Garnacha vines in Rioja Alta's Najerilla Valley, including those used in Bodegas Juan Carlos Sancha's viñedo singular Cerro La Isa.

Old vineyards are fragile – even more so when they come up against economic reality. Most winemakers believe that well-cared-for old vines, with their deep root systems (which have been known to go down 9m/27ft and more on the oldest vines), produce wines of more complexity and concentration. However, they also produce fewer grapes and make less wine. Financial support is available for growers as an incentive to keep their old vines, but these precious, twisted old plants will always be under threat simply because they bring in less money. There's a reason that most vineyards are pulled out and replanted before they reach 40 years old...

Another argument – besides the debate over whether old vines make better wine – is that the preservation of old vines safeguards the wider environment and helps protect traditional methods and ways of life in rural areas.

In Rioja Alavesa, Bodegas Luis Cañas is undertaking a long-term research project on varietal recovery. Very old vineyards around Villabuena de Álava were planted with mixed varieties, and having escaped modernization and replanting, they are a source of diverse genetic material. This research is an ethical duty, the bodega says, to protect genetic diversity in the grapevines of Rioja, which adds value to the region. The project will also assure the typicity of their family wines, including the Luis Cañas Viñas Viejas, a white field blend from vines more than 50 years old.

Bodega Bideona produces wines from more than 300 small parcels of old vines around the same village of Villabuena de Álava. Cofrades Bideona is a field blend made from a tiny plot planted in 1945, and it includes more than eight grape varieties.

Luis Cañas, Rioja Alavesa

Ploughing by mule reduces compaction of soils in old vineyards

Where to find old vines in Rioja

Several bodegas offer tours of their old vineyards, including Bideona (bideona.wine), just south of Samaniego, Bodegas Roda (roda.es) and Luis Cañas (luiscanas.com). For more information on how to visit bodegas, see The best Rioja bodegas for tours and tastings, page 110.

Rioja's greatest wines

Ask a wine critic which wine offers the best value in the world, and invariably the answer is 'Rioja'. They may be talking about delicious entry-level crianza that retails for little more than a pint of beer in a London pub, or they may mean cultish cuvées that cost hundreds of pounds a bottle. But the fact is that, at every level, Rioja punches way above its weight in terms of quality-to-price ratio.

Wine Fandango, Logroño

Below are 12 of Rioja's greatest wines (there are many more). Whether you seek out the £300 Viña Tondonia, the even more expensive Mágico from Bodegas Sierra Cantabria, or Benjamin Romeo's £50 Contador, each of these wines has earned its reputation.

López de Heredia
Viña Tondonia Reserva Blanco

Founded in 1877 and still family-run today, López de Heredia was Haro's first bodega and the third in Rioja. Viña Tondonia is the name of a 100-hectare (247-acre) vineyard on the right bank of the River Ebro, used to make López de Heredia's top wines. The first reserva wine was bottled in 1890, with grandes reservas being made from exceptional vintages, including 1968, 1970, 1973 and 1976.

As well as long-aged reds, Viña Tondonia is renowned for its Reserva Blanco, an oak-aged white Rioja that can last for decades in bottle – at least as long as the red wines. It's one of the world's most highly sought-after white wines, with prices in the region of £300. Look out for the 2004, 2001 and 1996 vintages.

Marqués de Murrieta
Castillo Ygay Gran Reserva Especial Blanco

Marqués de Murrieta is a pioneering Rioja winery, founded in 1852. Its Castillo Ygay Gran Reserva Especial is a blend of Tempranillo aged in American oak and Mazuelo aged in French oak. Grapes are sourced from La Plana vineyard, which was planted in 1950 at the highest elevation on the estate in Rioja Alta.

Castillo Ygay Gran Reserva Especial Blanco is one of Spain's most famous white wines. The latest release, 1986, is made from 97% Viura with 3% Malvasia and was bottled in 2014 after an astonishing 21 years in American oak barrels and a further 67 months in concrete tanks. Concentrated and creamy, with aromas of citrus and exotic fruits, it is quite unlike any other white wine. Previous vintages include 1970 and 1946. Expect to pay £600 and upwards. (See The best Rioja bodegas for tours and tastings, page 110.)

La Rioja Alta
Gran Reserva 890

La Rioja Alta was founded in 1890, and the Reserva 1890 was its first release. The 890 (the '1' was dropped to avoid confusion with the vintage; the same happened with the 904, launched in 1904) is a classic Rioja with long ageing potential and the pinnacle of La Rioja Alta's range.

Tempranillo is fermented with a small amount of Mazuelo before being blended with Graciano and aged in barrel for six years. 2010 was the first vintage made since the acclaimed 2005, and it has received perfect 100-point scores. Price: from £130 per bottle. (See The best Rioja bodegas for tours and tastings, page 110.)

Artadi
Viña El Pisón

Juan Carlos López de Lacalle built
Artadi in 1985, when he acquired
a former cooperative of growers
with well-situated vineyards. The
winery remains focused on tradition
and respect for nature, farming
organically in the heart of the
Sonsierra Alavesa region.

The mild-mannered López
de Lacalle is famous for breaking
away from the Rioja appellation
in protest at not being allowed to
put vineyard names on his bottles.
Indeed, he was one of the early champions of the change in
the regulations that came in 2019 (see Rioja classifications
and how they work, page 26). Starting with the 2014 vintage,
Artadi wines have been labelled as Álava table wines. The
much-fêted Viña El Pisón is a single-vineyard wine from the
family vineyard, a 2.4-hectare (5.9-acre) plot planted in 1945
in Laguardia. These are some of the most renowned wines in
Rioja: the 2015, 2016 and 2018 have all been widely praised
and retail for upwards of £200.

CVNE
Viña Real Gran Reserva Especial

Founded in 1879, CVNE (Compañía
Vinícola del Norte de España,
always abbreviated and pronounced
'coo-neh') is another of the handful
of noble old bodegas that cluster
around the station at Haro in Rioja
Alta. It has a stable of famous
wines – the long-lived Imperial, Viña

Real, Monopole and Contino, among them – and its modest owner Victor Urrutia says matter-of-factly that CVNE is the most important winery in Spain.

Viña Real is named after the nearby Camino Real (royal highway) in Rioja Alavesa; the winery was an early champion of the vineyards in this region. Dating back to 1920, Viña Real Gran Reserva Especial uses grapes from small plots at altitudes of 500–650m (1,640–2,130ft). Aged for 24 months in new oak, both American and French, the 2015 vintage can be found for under £30. (See The best Rioja bodegas for tours and tastings, page 110.)

Bodegas Roda
Cirsion

The newest of the small band of Haro wineries by almost a century, Roda was founded in 1987 by Carmen Daurella and Mario Rotllant. Their idea was to combine an old-fashioned, artisan approach with New World sensibilities: reduce yield, pay attention to detail in the winery, use as many old bush vines as they could find, and use only natural yeasts in fermentation – all common practice now, but radical in the 1980s. Cirsion is Roda's top wine. Made only in the best years and only from the best parcels of Tempranillo, with about 10% Graciano, it's beautifully alive and fresh and is capable of ageing for decades. Expect to pay at least £150 a bottle. (See The best Rioja bodegas for tours and tastings, page 110.)

Lanzaga
Las Beatas

Beloved by sommeliers, Lanzaga is somewhat of a cult winery known for the elegance of its wines. After years of gaining winemaking experience in Bordeaux and Burgundy, Telmo Rodríguez returned home to Rioja and started buying old vineyards around the village of Lanciego in Rioja Alavesa.

Las Beatas is a single-vineyard wine from a small, 1.9-hectare (4.7-acre) plot in the foothills of the Sierra de Cantabria. Rodríguez discloses limited winemaking information but reveals that the plot has eight, or possibly nine, local varieties. Production is very limited, with prices starting at £150.

Palacios Remondo
Quiñón de Valmira

Álvaro Palacios is Spanish winemaking royalty. After helping to put Priorat and Bierzo on the fine-wine map, he returned to his family estate in Rioja Oriental and made it part of the new wave of Rioja producers.

Palacios Remondo calls on its high-altitude, old Garnacha vines to make wines with real depth and concentration. Quiñón de Valmira is a single-vineyard Garnacha with 10% other traditional varieties, grown above 600m (1,970ft) in Alfaro. Prices easily top £200 per bottle.

Viñedos y Bodegas Sierra Cantabria
Mágico

This is a hugely expensive, ripe and heady cultish wine from a family winery run with the conviction that, above all, they are grape growers. Mágico is produced from a single vineyard, El Vardallo, just 1.18 hectares (2.9 acres) in size, in San Vicente de la Sonsierra in Rioja Alta. The vineyard was planted in the 2000s with a majority of Tempranillo (65%), 20% Garnacha and a selection of other varieties, both red and white.

In the excellent 2010 vintage, the family wondered if making a field blend with all the varieties from El Vardallo would result in a remarkable wine, and the first vintage of Mágico was made. Bottles of the 2019 currently sell for around £500.

Marqués de Riscal
Barón de Chirel

Now as famous for its Frank Gehry-designed hotel as for its wines, Marqués de Riscal was founded in 1858 by Don Guillermo Hurtado de Amézaga; four years later, it became the first bodega to bottle a Rioja wine. Barón de Chirel was first produced in 1986, establishing itself as a classic in what was then the new era of Rioja wines. Made from vines over 80 years old, it is a blend of Tempranillo with other unnamed varieties, aged in French oak for 16 months. Riscal is a noble bodega with one of the wine world's most dramatic buildings, but you might find wines like the Chirel quite old-fashioned in style. Bottles of the 2019 can be picked up for around £100. (See The best Rioja bodegas for tours and tastings, page 110.)

Bodegas Muga
Aro

Another venerable old bodega, Muga was founded in 1932 by Isaac Muga and Aurora Caño in the Barrio de la Estación in Haro; it is now run by their three sons and daughter. It's a huge operation, controlling some 300 hectares (750 acres) and with 17,000 barrels in its cellars, and it is also one of Rioja's most traditional wineries. As well as having vineyards in different sites around Haro, Muga makes all its own barrels in a fully functioning cooperage.

Muga's most famous wine is Prado Enea, but its most expensive is the dense and concentrated Aro (a reference to one of the old names for Haro). Made from Tempranillo and Graciano, Aro is aged for 18 months in new French oak barrels. Bottles of the limited 2019 production are selling for around £200; they will age for at least a couple of decades. (See The best Rioja bodegas for tours and tastings, page 110.)

Bodega Contador
La Cueva del Contador

In 1995, Benjamín Romeo – who was winemaker at Artadi (see page 39) for 15 years – bought a centuries-old cellar under the castle of San Vicente de la Sonsierra. It is here that, in 1999, he made his first vintage of La Cueva del Contador, produced from vineyards of Tempranillo up to 80 years old, vinified in his rock-hewn cellar and in his parents' garage. In 2008 he completed a new winery just below the village of San Vicente, an ultra-sustainable, low-level, plant-roofed building whose concrete walls are designed to gradually become coated with dust and so 'meld with the earth from which they came', as he says on his website. Romeo's wines, aged for 18 months in French oak, have been lauded as 'incredibly exuberant yet perfectly structured'; several of his cuvées have been given perfect 100-point scores. New vintages can be found for around £50.

Artadi's bush-grown vines looking towards the Sierra de Toloño, Sonsierra, Rioja Alavesa

Geography

The rolling vineyards of Rioja are a majestic sight to behold – whether in the first bloom of spring or in autumn's dappled post-harvest copper and gold. Centred on the capital city of Logroño, Rioja is a vast region, stretching around 100km (60 miles) from west to east and 50km (30 miles) north to south.

Briñas, Rioja Alta, with the Sierra de Toloño in the distance

The mighty Ebro, Spain's longest river, runs through Rioja as it makes its way from its source in the northerly Cantabrian Mountains towards the southeast and the Mediterranean coast – a journey of nearly 1,000km (620 miles). The Ebro has huge significance for Rioja, its tributaries creating the seven river valleys that bisect the region, producing a variety of soil types and aspects.

Across these seven valleys sit three subregions: Rioja Alta, Rioja Alavesa and Rioja Oriental (formerly known as Rioja Baja). Each has a slightly different mixture of soil types and altitude, which is why many traditional wineries still choose to blend grapes from across the three regions to create an interpretation of Rioja as a whole.

Broadly speaking, Rioja has a warm, continental climate, though this differs between subregions according to their altitude and respective proximity to the Atlantic and the Mediterranean oceans. The Sierra de Cantabria mountain range to the north shields Rioja from cold Atlantic winds: without this protection, temperatures would be too low for quality viticulture, and Rioja would never have become the famous wine region it is today.

Some of the differences are also political because Rioja falls into three political regions, or *autonomías*: La Rioja, País Vasco (Basque Country, or Euskadi in Basque, of which Rioja Alavesa is part) and Navarra. Note that the wine region is always known as Rioja, without the definite article.

In terms of size, Rioja is by far the biggest wine region in Spain. With its 65,000 hectares of vineyards accounting for 7% of all plantings in the country and about 300 million litres of wine produced per year, it dwarfs Priorat (2,000ha), Rías Baixas (4,200ha) and Ribera del Duero (26,000ha), to name just three regions that are Rioja's equals in terms of reputation. As a brand, Rioja is so well known worldwide that it tends to eclipse other Spanish wine regions. Ask the average person where a famous wine like Torres Gran Sangre de Toro comes from, and chances are they will answer Rioja. (It is actually from Penedès, in the northeast.)

Rioja is a quintessentially northern Spanish region. It has an Atlantic climate, with rain year round and distinctly chilly winters. Some wine producers – notably Roda – label the style of each vintage Atlantic or Mediterranean, depending on the northern or southern influence of the weather, but Rioja has always looked north rather than south. After all, the region's reputation and wealth come from a long association with Bordeaux (see History, page 17). Riojan food is hearty, often characterized by warming stews and green vegetables. An annual festival in Calahorra celebrates Riojan produce: tomatoes, peppers, asparagus, artichokes, mushrooms, lettuce and chard are all abundant here.

This is a region with an ancient history. The landscape is dotted with 2,000-year-old monuments and Romanesque churches; villages cluster around hilltop forts; and in the vineyards you can find pre-Roman stone winemaking vessels. Yet there is a radicalism at the heart of Rioja. Winemaking rules have been radically overhauled (see Rioja classifications, page 26); wineries such as Riscal, Baigorri and Ysios have commissioned avant-garde architects to design their breathtakingly modern headquarters; and this part of Spain has some of the most exciting restaurants in Europe (see dining sections, page 134).

Rioja Alavesa

North of the River Ebro, in the province of Álava, Rioja Alavesa has a recorded history that goes back 4,700 years, as revealed by archaeological remains discovered in the beautiful village of Elvillar. Another significant location in Rioja Alavesa is the settlement of La Hoya, in the district of Laguardia, which dates back to 1500 BCE and is one of the most important Bronze Age sites in the Basque Country.

Rioja Alavesa consists of 15 municipalities and 24 charming villages, and it is home to many of the smaller, family-owned wineries, such as Artuke and Sierra de Toloño. These lesser-known gems are increasingly recognized following the DOCa rule changes that permit individual village and vineyard names to be displayed on labels.

The climate of Alavesa is continental, with warm summers, cold winters and substantial winter rainfall. The soils are mostly clay-limestone, a much-prized soil type that can produce wines with low alcohol levels, high levels of natural acidity, and fresh, clean flavours. Most vineyards here are planted at altitudes of 400–1,200m

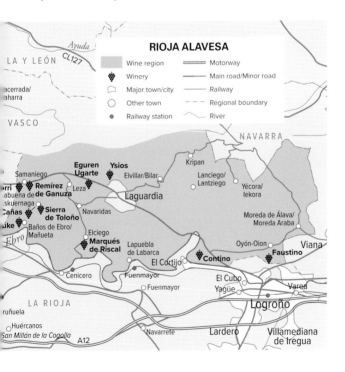

(1,310ft–3,940ft) above sea level; this is crucial, because the higher you go, the greater the difference between day- and night-time temperatures. A wide diurnal range – hot days and chilly nights – is ideal for winemakers, with cool nights resulting in grapes with higher acidity, which gives them freshness. Many of the famously terroir-driven projects in Rioja, such as Artadi and Artuke, began in Rioja Alavesa. Tempranillo accounts for 80% of all plantings, and it is supported by its traditional blending partners: Garnacha, Mazuelo and Graciano.

With about 13,500 hectares of vines, Rioja Alavesa is the smallest of Rioja's three subregions, but it is renowned for its Tempranillo, which has a vivid colour and refreshing acidity. Part of the region's success is due to the protection offered by the nearby Sierra de Cantabria, which shelters the vineyards from the cool winds and allows heat to be retained long into the autumn harvest; this is particularly important because of its proximity to the cooling effects of the Atlantic Ocean. It produces many of Rioja's most celebrated white wines, typically made from Viura, Malvasia and Garnacha Blanca. There are currently 119 wineries in Rioja Alavesa, including famous names such as Marqués de Riscal, Remelluri and Remírez de Ganuza.

While many traditional wineries are still focused predominantly on multiregional blends, vino de zona wines are on the rise, highlighting the unique nature of Rioja Alavesa. Some of the most famous single vineyards are in this northerly subregion, and many of them have gained significant critical acclaim. Vineyards such as Artuke's La Condenada, whose wines have recently been heaped with praise, are testament to the way Rioja is exploring the full possibilities of its terroirs.

In many ways, Alavesa is the essence of Rioja, since here you can see the evolution of the region in perfect contrast. For example, in the village of Elciego, the titanium plates of Frank Gehry's Modernist masterpiece, the rippling exuberance of gleaming metal that is Marqués de Riscal, are seen alongside buildings that date back a thousand years.

Alavesa also has one of Rioja's most famous wine routes (see page 100), heading through charming villages – Labastida, Oyón, Villabuena, Elvillar, Baños de Ebro and Lapuebla de Labarca – each with its own cluster of bodegas. (See page 110 for those that can be visited.) The villages' close proximity to one another is a key factor in the popularity of the wine route. A particularly noteworthy destination is the regional capital Laguardia, a hilltop town of narrow cobblestone streets founded in the 10th century as a defensive point for the Kingdom of Navarra. This bustling town of 1,500 inhabitants is regarded as one of the most beautiful in Spain – an award it has won several years running.

Rioja Alta

Rioja Alta stretches south of the River Ebro, from Haro to Logroño. With 28,000 hectares of vineyards, it accounts for almost half the wine production of Rioja. The climate here is continental, slightly warmer than Rioja Alavesa but still with that crucial difference between day- and night-time temperatures. Alta also boasts some of Rioja's most arresting landscapes: the region is laced with hiking trails into the Toloño, Cantabria and La Demanda mountains, as well as dramatic paths along the Ebro.

Alta is home to some of the region's most famous wineries, particularly around the city of Haro. Here is the

Barrio de la Estación, where in the 19th century millions of litres of wine were exported north to Bordeaux (see History, page 17). Clustered around the Estación de Haro are CVNE, La Rioja Alta, Gómez Cruzado, Bilbaínas Viña Pomal, Muga and Roda. The greatest concentration of wineries over 100 years old is found here.

The soil types in Rioja Alta are varied, as you might expect from a large region, and they include the clay-limestone soils more typical of Rioja Alavesa, as well as clay soils high in sandstone and iron oxide. Tempranillo, Mazuelo and Viura dominate plantings, and while there are many small, terroir-led projects here, this is the home of traditional winemaking. Compared to those of Rioja Alavesa, the wines tend to be lighter, with a focus on elegance, acidity and ageing potential. Red wines often see extended barrel ageing, and the designated ageing requirements for crianza, reserva and gran reserva are often exceeded. The wines made by the Barrio de la Estación producers, in particular, are considered the benchmark for this style.

Despite its size, Rioja Alta is also home to some of the most terroir-focused producers, particularly those focusing on single-vineyard sites. Miguel Merino's acclaimed Quinta Cruz vineyard is a great example of how the very best of the terroirs here are being highlighted, while Javier Arizcuren's urban winery in Logroño showcases Rioja Alta at a smaller scale.

Rioja Alta is split into 77 different municipalities and includes some of the most famous winemaking villages: Briones, Fuenmayor and San Vicente de la Sonsierra. There are many and varied wine routes to take, through the villages of Ábalos, Briñas, Casalarreina, Cenicero,

Cuzcurrita de Río Tirón, Sajazarra, San Asensio, Tirgo, Torremontalbo, Gimileo and Ollauri (see Wine routes, page 100).

Rioja Oriental

Until 2017 Rioja Oriental was known as Rioja Baja, due to its lower altitude – it sits at around 300m (980ft) above sea level. The name was changed to avoid any implication of lower quality, which is a suggestion the region has struggled with for several decades. In truth, Rioja Oriental has a huge quality potential if the correct grapes are matched to the correct terroir – something that hasn't always been the case. Oriental is Rioja's biggest subregion, stretching some 80km (50 miles) from Logroño to the village of Alfaro in the southeastern corner.

There are a couple of crucial differences between Oriental and the other two regions. Oriental is warmer and drier; it is lower in altitude and nearer to the Mediterranean, and average summer temperatures here can be in the mid-30s C (mid-90s F). Historically, Oriental has attracted less attention than its neighbours, but that is changing fast, thanks to both the growing popularity of Garnacha and the pioneering efforts of a powerhouse called Álvaro Palacios, who made his name in Priorat in the 1990s and returned to Rioja to revitalize his family winery in Alfaro. Bodegas Palacios Remondo dominates this part of Rioja, and it has become synonymous with the finest wines of the region.

The soils here are mostly sandy and alluvial, which together with the warm, dry climate make it the ideal terroir for Garnacha, which is far more prevalent than Tempranillo. The wines of this area tend to be riper and richer, with higher alcohol levels – this part of Rioja has long been a source of grapes and wine to add heft and spice to traditional blends. While Palacios has concentrated on Tempranillo in Alfaro (he's also a champion of Garnacha), there are several important projects dedicated to expressing the individual nature of Rioja Oriental. Wineries such as Vinos en Voz Baja and Arizcuren are leading the way here, their hallmark style a fresh, vibrant Garnacha.

Historically, with the climatic conditions and soil types suiting it perfectly, Garnacha was planted even more broadly. Unfortunately, many growers and producers pulled up the vineyards and replanted with the more marketable Tempranillo in the 1980s and 1990s. The warm climate, combined with lower rainfall, often causes drought conditions – something that suits the old Garnacha bush vines. Not only that, but there is a general renaissance of Garnacha across Spain; the fact that Rioja Oriental has some of the oldest vineyards in the country means that you'll hear far more about this part of Rioja in the coming years.

The regional capital of Rioja Oriental, Logroño sits close to the border with Alta. This lively city is Rioja's tourist heart and gastronomic capital. Many of Rioja's finest restaurants can be found here, as well as the famous pintxo (tapas) destinations of Calle del Laurel and the surrounding streets.

The serrated mass of El León Dormido (the Sleeping Lion) stands guard over Elvilar, Rioja Alta

Wine styles

Even for those who know it well, the sheer
diversity of wine in Rioja can be bewildering.
A varied landscape across the main subregions,
hundreds of vineyards and dozens of villages
have made for a kaleidoscopic array of wines
and a multitude of styles.

It was not always this way; decades of steady evolution have led to a recent period of accelerated change, with the 21st century seeing many exciting new projects and developments. Today, Rioja is one of Spain's most fashionable wine regions, in part due to initiatives that showcase the changes being made. Small groups of producers, like the eight that make up the Rioja 'n' Roll collective, are key drivers in this, highlighting wines made in specific villages and vineyards. Indeed, many of the newer producers style their wines more similarly to those of Burgundy than to traditional Rioja, and the international market is responding accordingly. It's no longer uncommon to find what were historically considered niche wines in bars in Tokyo, New York and London.

Joven

These entry-level wines, made with little or no oak ageing, are typically fresh, fruity and easy to drink. Many are produced by carbonic maceration, a process (traditional in Beaujolais) that preserves fruit flavours and softens tannins.

Crianza

The basic level of the ageing pyramid, crianza wines are full-bodied, with notes of red and black fruit, though lighter than reserva wines.

Reserva

A year's extra ageing gives reserva wines more body and more defined tannins; along with the dark fruit flavours, you'll find notes of leather, cigar box, spice and dried fruit.

Gran reserva

These are only made in best vintages from the finest grapes, and they show aromatic intensity, sweetly spiced ripe red fruit, and rich, velvety tannins with toasted notes. The finest grandes reservas can be among the world's greatest wines.

It's important to note that none of these ageing classifications is better or worse than the other – for the most part, it's a stylistic choice, and some producers specialize in certain styles. For example, Artuke makes one of the most brilliant joven wines in the region, whereas La Rioja Alta specializes in wines that have undergone barrel ageing.

It is true, though, that many of the best producers create some of the most long-lived, complex and ethereal wines with barrel ageing, learning through decades of experience that a few years in oak are necessary to shape and soften the tannins, imparting sweet, spicy flavours to complement the fruit characters. Visitors to some of the famous wineries of Haro will be faced with vast cellars.

A small section of the 17,000-plus barrels at Muga

At any one time a bodega like Muga will have as many as 17,000 barrels.

Today, Rioja is home to a broad variety of wine styles, with traditional houses alongside a new generation of grape growers and winemakers. There is no one true expression of Rioja, and the region is all the richer for this vast diversity.

Traditional wines

'Traditional red' in Rioja nearly always means a blend of Tempranillo with Garnacha (known as Grenache in France and much of the New World), Mazuelo (Carignan) and Graciano, though other local varieties are also permitted. As well as giving the winemaker the broadest possible palette to work with when blending, this is also an insurance policy: the more varieties you have in different locations guarantees a crop, however difficult the growing season. In traditional Rioja, the supple tannin and red-fruited character of Tempranillo is usually complemented by the spice and heft of Garnacha, the acidity and herbal notes of Mazuelo, and the colour and perfume of Graciano.

The ageing system of traditional Rioja means that wines tend not to be fruit-forward but display the complex aromas of slow ageing after lengthy periods in both barrel and bottle. Fermentations will be fairly short and gentle, often in old wooden vats, with little colour and tannin extracted from the grapes. The wines are then aged in barrel, with many bodegas opting for the historically important, aromatic American oak. (This has a wider grain than French oak, so the wine extracts more flavours from the wood as it ages.)

Zaha Hadid's Tardis-like modernist tasting room at López de Heredia, Haro

Depending on their individual style, it's common to see producers surpassing the minimum ageing requirements for crianza, reserva and gran reserva – López de Heredia produces a crianza called Cubillo that is aged for three years in barrel and another three years in bottle before being released. At their best, the results are hauntingly beautiful, with a medley of desiccated red fruits, tobacco, mushrooms, earth and dried herbs.

MARQUÉS DE MURRIETA
GRAN RESERVA 2016

Desde 1852

RIOJA
DENOMINACIÓN DE ORIGEN CALIFICADA

Nº BOTELLA 00000 / 67.033

Acidity is as important as tannin for the lengthy ageing of these wines, and it can be surprisingly high, even in older bottles. Although they are often released at a comparatively mature age, these wines can age for decades.

The number of wineries still producing the traditional style is dwindling, although two iconic producers have stood the test of time and continue to champion long-lived, savoury wines. López de Heredia is perhaps the most famous, established in 1877 and witness to a different age entirely. Steeped in tradition as it is, López de Heredia is a perfect example of how Rioja marries ancient and modern: its famous old frontage is distinguished by a strikingly modernist tasting room designed by Zaha Hadid. Tondonia is its most famous vineyard and wine, but Cubillo and Bosconia are both compelling examples of Rioja at its most traditional.

López de Heredia's neighbour and rival for the throne is La Rioja Alta, producing a superb range of wines with comparable, lengthy ageing regimes and late releases. The intricate, powerful 890 and 904 gran reserva ranges are the

The barrel room at Viña Real, Rioja Alavesa

flagship bottlings, with the Ardanza Reserva perhaps the most recognisable. Both producers are classic examples of traditional Rioja, along with other wineries such as Urbina, Hermanos Peciña, Lagunilla and Marqués de Murrieta.

Traditional white wines

In Rioja, traditional white wines are dominated by the fresh, floral Viura grape blended with smaller quantities of Malvasia and Garnacha Blanca. They were made in a more rustic style – without the technology to keep temperatures cool and oxygen away from the wines, they tended to have a nutty, savoury character. Historically these wines made up just a tiny fraction of the overall production, and they were a niche consideration; however, since the turn of the century they have become increasingly sought-after to

the point of severe price inflation as supply outstrips demand. The result of long ageing and gentle oxidation is remarkable, with the fresh fruit flavours transforming into dried fruits, nuts and honey, while maintaining the fresh acidity that Viura is known for.

The standard-bearer for this style is, without a doubt, López de Heredia. As with this bodega's red wines, the whites are separated into age classifications and vineyard origins, with Gravonia the lightest and freshest of the wines, followed by Tondonia Reserva and Tondonia Gran Reserva. Rarer still is the legendary Castillo Ygay Blanco Gran Reserva made by Marqués de Murrieta; at the time of writing, the most recently released vintage of this wine is 1986. Due to the increased global demand for this style, more wineries are looking to the past to produce similar

Racking (changing barrels) at La Rioja Alta

offerings, such as CVNE's nutty, savoury Monopole Blanco Gran Reserva, and the delicious La Granja Nuestra Señora de Remelluri Rioja Blanco by Bodega Remelluri.

Modern wines

With the introduction of stainless steel and temperature-controlled fermentation in the 1970s, the styles of wine produced in Rioja started to change, becoming fresher and more fruit-driven. The 1990s saw the ascendancy of wine regions such as Ribera del Duero and international critics who favoured powerful, heavy wines; as a result, many producers in Rioja shifted again to emulate this style.

Modern Rioja reds are still mostly dominated by Tempranillo, but it's also now common to find wines made entirely of Mazuelo, Garnacha or other permitted varieties, often reflecting individual subregions or vineyards. Field blends are also increasingly common, often with white grapes co-planted with red varieties. It is an exciting time for Rioja as a new generation of winemakers seeks to unlock its various identities across not only the three subregions but also the winemaking villages.

With many producers taking advantage of the new regulations (see Rioja classifications, page 26) and exploring individual terroirs, whether village or single vineyard, it's safe to say there is no single template for modern Rioja. The winemaker's aim is to interpret the land, to make wine that reflects its terroir. It is a balancing act. One winemaker might decide that their land produces Tempranillo that needs at least some time in oak; another will feel that concrete tanks produce the purest expression of their terroir. As a general rule, long oak ageing is less likely; and extractions – stirring the fermenting wine to get colour and tannin – tend to be gentle. Used oak barrels and concrete vats are preferred to new oak, with the intention of letting the wine itself shine. Most modern producers eschew the traditional labelling of crianza, reserva and gran reserva, focusing instead on highlighting the terroir, whether from a

village or a single vineyard. These producers are found across all subregions of Rioja and include rising stars such as Jose Gil, Artuke, Arizcuren, Miguel Merino and Sierra de Toloño.

'Modern classic' red wines

The modern classic reds sit somewhere between the traditional and modern. The grape blends are more likely to be traditional in nature – that is, dominated by Tempranillo and mixed with Garnacha and Mazuelo – and often from a selection of vineyards across subregions. However, lengthy ageing in the traditional style is uncommon, and there is often a mixture of oak types: new and old, American and French. The wines tend to be fuller-bodied yet balanced, with rich fruit flavours and noticeable oak. Look out for the wines of Muga (see The best Rioja bodegas for tours and tastings, page 110), with its range produced at the Estación de Haro. Roda and Baigorri also make good examples of the modern classic style.

Modern white wines

This is a type of wine that is becoming ever more popular in Rioja. The introduction of temperature-controlled fermentation in the 1970s opened up new winemaking possibilities for white varieties; since then, techniques have been honed and styles have changed radically. As recently as the early 2000s, the heavily oaked expression was prevalent in Rioja whites, but it is now easy to find light, fresh styles. Viura still dominates plantings and is often blended with Malvasia and Garnacha Blanca, but there is also space for grapes from further afield, with Verdejo, Chardonnay and Sauvignon Blanc also permitted varieties in Rioja DOCa.

With the cooling influence of the Atlantic Ocean to the north of Rioja, the highest-altitude plantings in Rioja Alta and Rioja Alavesa are increasingly prized for their white-wine production. The lightest and freshest of wines rarely see new oak and are valued for their fresh fruit characters, florality and brisk acidity. Producers such as Palacios Remondo and Laventura make excellent, benchmark examples of this fresher, lighter style of white Rioja. Importantly, however, there has been a quiet revolution for more ambitious white wines in Rioja since around 2010 or so. These new-wave wines, mostly led by old-vine Viura, have a depth and texture to them that have captured the attention of the wine-drinking world. The likes of Abel Mendoza, Nivarius, Carlos Sánchez and Cosme Palacio are all producing superb examples, while single-variety wines made from Garnacha Blanca, Malvasia and even Maturana Blanca are also on the rise.

The tank room at Viñedos de Aldeanueva, Rioja Oriental

Grape varieties

Rioja has long been the homeland of Tempranillo, the mainstay of the region's greatest wines. But as the region evolves, a new generation of ambitious and iconoclastic winemakers is enthusiastically experimenting with all 14 of the grape varieties (five red and nine white) that are permitted here, as well as Merlot, Chardonnay and other international varieties.

Harvesting by hand is still widespread in Rioja

Red grapes

Tempranillo The most famous of all the Rioja grapes and considered indigenous to the region, Tempranillo is the most widely planted grape variety in Rioja – and for good reason. It ripens early (hence the name: *temprano* means early), it thrives on various soils at different altitudes, and it works in a range of winemaking styles.

As a varietal wine – that is, one made from a single grape variety – Tempranillo traditionally shows aromas of ripe red fruits, herbs and violets; in warmer microclimates, it gives flavours of blackcurrant and blackberry – and sometimes blueberry. More commonly, however, you will find Tempranillo as the main element of a blend. Testament to its versatility and longevity, the variety has more than 70 synonyms across Spain: in neighbouring Ribera del Duero, it's called Tinto Fino or Tinta del País; in La Mancha and Valdepeñas, it's Cencibel; in Catalonia, it's Ojo de Liebre and Ull de Llebre; and in Madrid, Tinta de Toro.

As much as 85% of red wine in Rioja is produced from Tempranillo, either as a varietal wine or as a blend. In a blend, it plays the same role that Cabernet Sauvignon does in Bordeaux: it provides much of the texture and structure to the wine, while also giving it its long ageing potential. Garnacha is often used to fill the mid-palate and add weight, while Graciano adds acidity and floral characters. Importantly, Tempranillo works well with oak, too, absorbing flavour without being overwhelmed. Aged Tempranillo, in particular, gives delightful aromas of dried tobacco, cedar, sweet spices and undergrowth.

Tempranillo has so many different styles that it is hard to pin down its character. In the higher-altitude vineyards of Rioja Alavesa, wines made from the grape tend to be deeper in colour and slightly fuller in body, with darker fruit aromas and more extract. In Rioja Alta, the wines are lighter in body and higher in acidity, erring towards a more elegant, traditional style. In Rioja Oriental, Tempranillo tends to be fuller, richer and more red-fruited. Many wineries blend together not just different grapes but also a mixture of Tempranillo wines of different styles to bring out the individual strengths of all these various characteristics.

Garnacha Historically, Garnacha has been a workhorse grape in Spain. In Rioja, it was used to plump up Tempranillo blends, or it was vinified as a simple wine for the vineyard workers. Today, it is one of the most exciting prospects in the area, where its lightness of body and aromas of ripe red fruits, wild herbs and white pepper are very much in fashion. There are a great many old Garnacha vineyards, particularly in Rioja Oriental, that are increasingly being used to produce premium wine. High-quality Garnacha still brings a great deal to Tempranillo blends, increasing the alcohol level of the wine, softening the tannins and adding juicy fruit characters.

Garnacha

Increasingly, some of the most celebrated wines in Rioja are being produced from a majority of Garnacha. The grape has undergone something of a renaissance in Spain, and many of Rioja's smaller premium producers are making varietal wines from old vineyards at a very high quality level. The wines tend to be produced with whole-bunch fermentation and with little to no oak ageing, to allow the grape to shine through. Indeed, many winemakers consider Garnacha the ideal variety to show a sense of place, particularly when produced from older vines.

The sandy soils of Rioja Oriental are perfect for the production of Garnacha: the cool, loose soils allow for a slower maturation of the grape, lowering alcohol levels and leading to a better retention of acidity. However, there are many old Garnacha vineyards all across Rioja, with some vibrant, juicy examples being made even in the far north of Rioja Alavesa.

Graciano Although Graciano makes up only a small percentage of the vineyards of Rioja, its importance far outweighs its volume. It is highly prized for its dark colour, ripe tannins and vivid, floral character. Varietal Graciano wines are rare but are celebrated for their intensity and verve. Most Graciano grapes are used in minor roles in traditional blends, where they add complexity and concentration.

Mazuelo Better known as Carignan elsewhere in the world, Mazuelo is an increasingly important black grape in Rioja. Born of high-yielding vines, traditionally this variety was used to bulk up volumes and naturally improve levels of acidity in wines, but older Mazuelo vineyards are now producing some of the most exciting wines in Rioja.

Maturana Tinta A tiny percentage of Rioja's vineyards is planted with Maturana Tinta, a grape with links to France's Castets variety. It is black-fruited and often intensely herbal, and it can give soft, juicy wines of real interest when picked at full ripeness and handled sensitively.

Tempranillo Blanco

White grapes

Viura Also known as Macabeo across the rest of Spain, Viura is the most planted white grape variety in Rioja. It dominates across all white-wine styles, from traditional to modern. With its naturally high acidity, bright fruit and floral character, Viura is an ideal foundation for high-quality white wines, though when grown at high yields it has a neutral character. Importantly, it expresses itself well both as a fresh, fruity wine and as a more textural, barrel-fermented wine.

Malvasia One of the oldest known grape varieties around the world, Malvasia is also grown throughout Rioja, where it is called, unsurprisingly, Malvasia Riojana. It is rarely used to make still white wines, but it is instrumental in adding a floral lift to white blends.

Tempranillo Blanco An increasingly important white variety in Rioja, Tempranillo Blanco began life as a single vine mutation in 1988. It now accounts for 12% of white plantings across Rioja, and ever more winemakers are seeing its potential. It can produce wines with a soft texture, high levels of refreshing acidity and delicious flavours of tropical fruits and white flowers.

Other white grapes The rest of Rioja's white grapes are planted in tiny quantities. **Maturana Blanca** is the white mutation of Maturana Tinta and lays claim to being the oldest of Rioja's grape varieties. **Garnacha Blanca** is often included in blends but is increasingly used to produce varietal wines. Other grapes include the early-ripening **Turruntés**; more recently, **Verdejo, Chardonnay** and **Sauvignon Blanc** have been permitted by the DOCa.

The peaks of San Lorenzo from the village of Badarán, Rioja Alta

Wisps of autumn mist fade during harvest in Briñas, with the distinctive cupola of the Restaurante Portal de la Rioja in the background

How to read a Rioja label

Labelling regulations are strict. As we have set out in Classification (page 26), since 2017 bodegas are now allowed to classify their wines geographically – by region, village or vineyard. There are of course myriad different styles of label; modern Rioja produces some of the coolest artwork around, and the new regulations mean there are now many new ways of expressing the uniqueness of your wine. These two pages are a very brief guide to the most common terms found on traditional and modern Rioja labels.

Traditional label

Marqués de Murrieta: The name of the bodega, or winery

Gran Reserva: this confirms that the wine has been aged for at least 2 years in 225-litre oak barrels, and at least 3 years in bottle

Castillo Ygay: Ygay is Marqués de Murrieta's most famous vineyard, Castillo Ygay the manor house and winery

Cosecha 2012: Vintage 2012 – the year the grapes were harvested

Especial: this has no legal significance

Rioja Denominación de Origen Calificada: The DOCa standard is confirmation that this is a Rioja appellation wine. The wine conforms to regulations overseen by the regulatory body, the Consejo Regulador. DOCa is higher standard than the more common DO (Denominación de Origen)

Modern label

Queirón: The name of the bodega, or winery

Viñedo Singular: Single Vineyard. This is a legal term under the new Rioja regulations (see page 26) that guaranteees the wine comes from a single vineyard of vines more than 35 years old

El Arca: The name of the vineyard

Garnachas Antiguas: Old garnacha vines

0,89 Ha: the size of the vineyard in hectares (just over 2 acres)

2.016 BOT: The number of bottles that were made this year. Not a labelling requirement but an indication of how small the vineyard is

Barrio de Bodegas de Quel s.XVIII: no regulatory significance. Refers to the old wine quarter of the town of Quel in Rioja Oriental, where the cellars beneath the town date back to the c18th

Rioja Denominación de Origen Calificada: The DOCa standard is confirmation that this is a Rioja appellation wine. The wine conforms to regulations overseen by the regulatory body, the Consejo Regulador. DOCa is higher standard than the more common DO (Denominación de Origen)

HEREDEROS DEL
ARQUÉS DE RISCAL
ELCIEGO (ÁLAVA)
RIOJA
DENOMINACIÓN DE ORIGEN CALIFICADA
RESERVA

FUNDADA EN 1886
Gómez
Cruzado
HARO · RIOJA
Cosecha
RIOJA
DENOMINACIÓN DE ORIGEN CALIFICADA

RIOJA
Denominación de origen calificada

CIRSION

2019

BAI
GORRI

BLANCO FERMENTADO
EN BARRICA

Vino blanco elaborado con uvas de la
variedad viura seleccionada grano a
grano. La elaboración se desarrolla
en barrica de roble francés durante seis
meses sobre sus lías, otorgándole una

Aro
R I O J A
DENOMINACIÓN DE ORIGEN CALIFICADA
Muga

MONTE GATÚN
EMBOTELLADO EN ORIGEN
VIÑEDOS EN PROPIEDAD

arizcuren

*Garnacha, Mazuelo y Tempranillo,
variedades que unen pasado y futuro de nuestra tierra*

SERIE NUMERADA   0012 / 4.561

EVA DE LOBOS

Great Rioja vintages

Since 1925, Rioja's governing body, the Consejo Regulador, has been rating each vintage *excelente*, *muy buena*, *buena*, *normal* or *mediana*. This ranking should be taken with a pinch of salt (the last *mediana* was 1972, and there hasn't been anything less than a *buena* vintage since 1979), especially because it's hard to speak generally about a region as large and varied as Rioja. However, over the years certain vintages have shone, of course.

La Rioja Alta's Finca Las Cuevas in Rodenzo. Sheep are excellent organic mowers and fertilizers

Below are some of the legendary vintages of Rioja. These are years that have stood the test of time and show not only quality but a stylistic individuality. It is still possible to find bottles of even the rarest of these wines, particularly in restaurants that specialize in older vintages, such as Rekondo in San Sebastián, Kaia Kaipe in Getaria and El Celler de Can Roca in Girona.

1928 Europe was still reeling from the devastation of World War I and on the verge of a financial crisis, yet the conditions in Rioja were perfect for creating wines with longevity. Wines from Bilbaínas and Federico Paternina are still showing amazing vitality today.

1955 Only the fourth year to be rated *excelente*, 1955 would go on to become one of the most famous vintages of all time, with wines from López de Heredia, in particular, highly sought after.

1961 One of the greatest vintages of the century in Bordeaux, 1961 in Rioja is less stellar but considered to have produced some exceptional wines. The Tondonia Gran Reserva wines from López de Heredia, in both white and red, are currently at their apex some 60 years later.

1964 Hailed as the vintage of the century, 1964 is considered by many wine lovers to be Rioja's greatest. As with all older wines, there are hits and misses, but wines from producers such as Bodegas Riojanas (particularly its Monte Real bottling) are ethereal beauties.

1970 While not considered *excelente* at the time, 1970 has stood the test of time, and many of the wines have been thrilling to drink over the past 40–50 years. Producers such as Faustino, Martínez Bujanda and Salceda made stunning wines that are still drinking well today.

1982 An acclaimed year across much of Europe and especially in Rioja, 1982 is overshadowed by the excitement over Bordeaux's legendary vintage. A warm year produced bolder wines than the '64s, for example, and they have unfolded beautifully over time. Wines from Bodegas Faustino and Montecillo are particularly good.

1994 Often compared favourably to 1964, this was a year of triumph after three difficult vintages in the early 1990s. Occasionally, these moments of relief end up in exaggerated quality assessments, but 1994 has stood the test of time. La Rioja Alta in particular produced some truly stunning wines.

2001 The first of the great vintages of the 'modern' era of Rioja, this was also – unusually – a bumper year for yields, which does not always equal quality. Many bodegas released special-edition wines in 2001, and many of these are still at a very youthful stage almost a quarter of a century later.

A cooper at work at Muga. Barrel-making is a craft almost unchanged for 3,000 years

2010 Almost perfect growing conditions: a warm, dry summer and some rain in September. Many producers were delighted with the quality; wines slightly more robust and with higher alcohol than average. A number of bodegas produced stunning wines this year, with Muga, López de Heredia and CVNE being especially good.

2016 When wines are so young, it's difficult to judge how they will be 20 or 30 years later, when most top Rioja is opened. However, 2016 has all the makings of a classic, with fresh, well-defined and supple wines across a number of producers. Despite the warmer weather compared to 20 years earlier, these are fresh, vital offerings that are a great choice for long-term ageing. Not legendary yet...

Visiting Rioja

When is the best time to visit Rioja – in winter, when the vineyards are dusted with snow and woodsmoke scents the air; in the height of summer; or in spring, when the vines are budding and musicians tune their instruments for the first great festivals of the Rioja year?

The Dancers of Anguiano

In the winter, you'll find the bodegas quiet and empty. August, when temperatures can be ferocious, is also a tranquil time, before the bustle of harvest. For many wine lovers, autumn is the most rewarding time to visit. Harvest is under way, the vineyards are dotted with pickers, and the bodegas hum with activity. In the smaller towns and villages, queues of trucks and tractors rumble along the lanes, each laden with the fruit of an entire year's work in the vineyard.

Winter (December to February) Winter is one of the few times of the year when winemakers and their teams are not under enormous pressure. Consequently, you are more likely to be offered an extended visit at a bodega. The hearty food of northern Spain is also ideal for the wintry weather: slow-roasted lamb with a sappy, vibrant bottle of Tempranillo – heaven!

Spring (March to May) Spring is a time of celebration in Rioja. It's the start of the growing season and all that it promises: imminent warmer weather and the launch of regional festivals. The weather is typically fresh and breezy, and you should expect rain, particularly in March and April.

Summer (June to August) Summer in Rioja is magical. The blistering heat of the day gives way to long, mellow evenings. This is the most popular season for visitors, and the prices are at their highest. The tapas trail in Logroño (see page 149) is of particular interest, with visitors and locals alike eating and drinking late into the evening. It's also festival time. Be aware, though, that August can be quiet, with restaurants closed and winemakers on holiday.

Autumn (September to November) The vineyards are beautiful in the autumn months as the leaves turn from green to gold. Although the weather has cooled, September can still be very warm. This is also harvest season – the busiest time of year for the bodegas. You won't be able to get the winemaker's attention, but it's fascinating to be in a wine region when every town, every vineyard and every bodega is a frantic bustle of activity as the grapes come in. Remember also that school holidays are over by now, and most of the regional festivals have passed.

Festivals

The most exciting – and most unmissable – of the Rioja festivals is the legendary Battle of Wine at the end of June, but there are many others. Some are based around religious events and dates; others are rooted in nature, wine and the growing season. There are music festivals, food festivals, wine festivals and festivals that simply celebrate life. Most festivals take place between early spring and the end of autumn, with the majority concentrated around late spring and summer.

The following list covers some of the more notable events, but it certainly isn't exhaustive. Villages all across Rioja put on their own events – fascinating hangovers from the religious wars, like the Procession of the Hundred Maidens of Sorzano on the third Sunday in May; oddities like

La Batalla del Vino, Haro

the Medieval Days of Briones in June; or ancient rites like the Holy Week processions in San Vicente de la Sonsierra.

Álava Olive Oil Festival

Moreda de Álava, Lanciego and Oyón (late March)
This started as a small, local affair but is gradually growing in size and importance. Spain is the world's largest producer of olive oil, and the much-favoured Arróniz olive is unique to Rioja. Expect to attend various guided tastings, visit olive oil farms and see how it's made, see plenty of street entertainment and, of course, eat plenty of local tapas and pintxos that highlight the importance of olive oil in Spanish cooking.

Celebration of Tapas and Pintxos Calahorra (April)

Dozens of bars and restaurants showcase their best creations, all washed down with ample amounts of wine, and prizes are awarded for the best tapas. *Concursos de tapas* – tapas competitions – take place all over Spain and are incredibly popular; people come from outlying towns to taste the wares and cast their votes.

The Dancers of Anguiano (May, July, September)

One of the most popular festivals in Rioja, Los Danzadores de Anguiano could equally be called 'The Dance of the Stilts'. The town of Anguiano honours patron saint Mary Magdalene through this ancient ritual, a dramatic spectacle in which eight men on stilts, wearing multicoloured waistcoats and voluminous yellow skirts, dance and swirl through the town rattling their castanets. The origins of the dance are thought to be pagan, a tribute to the sun – hence the bright yellow skirts.

The Festival of St John (24 June)

San Juan, or John the Baptist, is the patron saint of Catalonia and is celebrated across Spain on 24 June. Families take the day off to enjoy a day in the countryside, but it's the evening when the party really gets going. Bonfires are lit in the streets and in the town squares (on the preceding night in many places), with the entire population out celebrating the coming of summer. This is the beginning of the great festival that culminates in the Battle of Wine on 29 June.

Haro Wine Festival, the Feast of St Peter and the Battle of Wine Haro (29 June)

La Batalla del Vino is one of the great Spanish festivals, on a par with the Sanfermines of Pamplona for excitement and spectacle (although there are no bulls involved). The festivities begin on 23 June, but the party proper starts on the 28th, culminating in a procession up to the Hermitage of San Felices de Bilibio on the morning of the 29th (La Fiesta de San Pedro), where about a thousand revellers pour, squirt and generally douse each other with wine. The party then descends to Haro and continues. A memorable occasion.

The Bread and Cheese Festival in Quel
Rioja Oriental (6 August)

Las Fiestas del Pan y el Queso de Quel is one of the oldest festivals in Spain, held annually in the charming village of Quel since 1479. At 9am on 6 August, a bugle sounds and the parade begins. Bagpipers play their way through the streets, followed by the procession of the image of the patron saint of Quel. What makes this event unique is the throwing of bread and cheese from the chapel balcony at 11:15am, commemorating how lunch was provided in the 17th century to those following the procession.

The Day of La Rioja Logroño (9 September)

El Día de La Rioja celebrates the region's autonomy, which was granted in 1982. The day begins with a parade through the streets of Logroño to the town hall square, and this is where the festivities begin in earnest. Activities are held all around the city, and beautiful decorations are put in place. The residents take the opportunity to express pride in their city and to showcase the traditions of the region.

The Festival of St Matthew Logroño (mid-September)
A week-long harvest celebration, Las Fiestas de San Mateo starts on the Saturday before 21 September. The event dates back to the 12th century and sees the streets of Logroño filled with music and dancing. Huge meals are served, and revellers quaff *zurracapote* (a local drink similar to sangria), attend wine tastings and watch sports. The festival concludes with the *quema de la cuba*, the burning of a barrel set alight by a string of fireworks, while a brass band plays.

Las Fiestas de San Mateo

Top 5 things to do in Rioja

1. Visit the wineries of the Haro Station Roda, Muga, CVNE, Gómez Cruzado, Bilbaínas Viña Pomal and La Rioja Alta are the six current members of the Barrio de la Estación de Haro. So much of Rioja's wine past began around this historic train station (see History, page 14), and five of these wineries were the pioneers, founded to feed France's insatiable need for wine after its vineyards were decimated by phylloxera. (Roda was founded 100 years after the rest but is equally important.) See The best Rioja bodegas for tours and tastings, page 110.

2. The Logroño tapas trail This is a Rioja rite of passage. On Calle del Laurel and the blocks surrounding it, you can find more than 50 tapas and pintxo bars. Choose a specific bar for specialist tapas, or simply drift from one to the next, having a tapa and a glass of wine in each. See Where to find the best tapas in Logroño, page 149.

3. Stop in at Marqués de Riscal and Ysios Make sure you visit the town of Elciego, if only to see the shimmering waves of silver, gold and purple titanium that clad the extraordinary Riscal building, the work of Frank Gehry (of Guggenheim Bilbao fame). As the sun sets, the iridescent plates reflect and refract the light and change colour. Santiago Calatrava's Bodegas Ysios, a couple of kilometres north of Laguardia, is equally impressive, its swooping, jagged curves perfectly echoing the stark sierra behind. Both wineries are open to the public and have drop-in wine bars.

4. Visit the Vivanco Museum of Wine Culture There are many museums worth visiting in Rioja, but this one is a must-see. Not only can you combine it with a trip to Bodega Vivanco itself, but the museum covers 8,000 years of wine

The serrated edge of the roof of Ysios

culture, bringing together many of the elements of Rioja. Various aspects of wine culture are explored – from its history, to the modern day – and there is also a collection of thousands of corkscrews. vivancoculturadevino.es

5. Head out to the villages If you are short on time, it's natural to gravitate towards Haro and Logroño as the major destinations. On longer trips, though, do visit the winemaking villages dotted around the region. Alfaro in the far east of Rioja Oriental, Samaniego, Cenicero, Briones and Elciego are all ancient towns with medieval churches, lovely secluded squares and narrow, winding streets. Many are also home to the most famous names in wine, but you don't even need a reason to go: simply wandering around these medieval villages is an afternoon in itself. See Wine Routes page 100.

The towns and villages of Rioja

From the hubbub of Logroño, to the quiet of Elciego or San Vicente de la Sonsierra, Rioja's towns and villages are typical of this part of Spain: working communities that welcome visitors but have lost none of their character to the rise of mass tourism.

Labastida, Rioja Alavesa

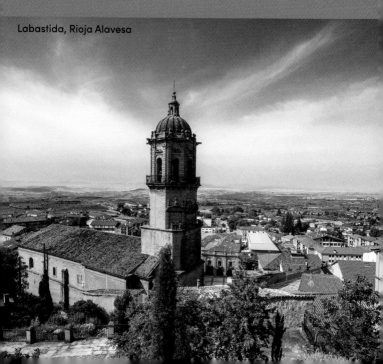

Haro

A town with a rich history, Haro has a cluster of some of Rioja's most famous wineries around its railway station, and the labyrinthine cobbled streets of its picturesque old quarter are lined with restaurants and upscale hotels. Although it is not as central as Logroño, Haro is still a good base for exploring the wineries of Rioja Alta. See Where to eat and drink in Haro (page 155).

Logroño

The region's capital, Logroño is a charming town that is geographically in the centre of Rioja, making it an ideal spot for a weekend visit. It is famous for the tapas bars on and around Calle del Laurel, and an evening spent wandering from bar to bar is a rite of passage for anyone who loves the Spanish way of life. Logroño is also an important stop on the Camino de Santiago, so you are likely to see dozens of pilgrims carrying wooden staffs and shells. The splendid cathedral of Santa María de la Redonda is just one of the fine old buildings standing in the old town. See Where to find the best tapas in Logroño (page 149).

Labastida

This beautiful walled town in Rioja Alavesa is notable for being the birthplace of 18th-century winemaker Manuel Quintano (see History, page 14). It is also home to Bodegas y Viñedos Labastida, one of the best cooperatives in northern Spain, and Granja Nuestra Señora de Remelluri. Originally a 14th-century monastery, Remelluri is now run by Telmo Rodríguez, one of Rioja's most prominent winemakers. Other places of interest in the area include the ruined monastery of Santa María de Toloño.

Elciego

In bustling Elciego – as in much of Rioja – ultra-modern bodegas stand alongside centuries-old monuments; indeed, some of the buildings in this most typical of Riojan villages date back a thousand years. This is a town where families promenade in the Plaza Mayor and old men gossip on benches as tractors rumble by. If you stay at the Hotel Marqués de Riscal, you can gaze past its gleaming titanium plates to see the twin towers of the 16th-century Church of San Andrés looming over the village. See Wine routes (page 100).

Briones

Briones

Just 9km (5.5 miles) southeast of Haro, surrounded by vineyards, is the medieval hilltop town of Briones. The Church of the Assumption, with its magnificent baroque tower, stands above a cluster of ochre and white, red-tiled houses, and in front of the church is a plaza ringed by tall buildings with classic glass balconies. Eat at Dinastía Vivanco (see Best restaurants in the bodegas, page 140), and stay at the gorgeous Hotel Santa María Briones (see Best wine hotels, page 122). Nearby wineries that are open to visitors include Betolaza (betolaza.es).

San Vicente de la Sonsierra

The 10th-century hilltop town of San Vicente, surrounded by ancient monuments (see Wine routes, page 100), is just a seven-minute drive northeast of Briones. At 500m (1,640ft) above sea level, it offers stunning views across Rioja. Wine lovers should be sure to visit Bodegas Abeica (bodegasabeica.com), which runs guided tours to its traditional stone *lagares*.

Laguardia

Stay in the exceptionally pretty Laguardia, and you'll be just a few minutes' drive from bodegas Ysios and Marqués de Riscal. Cars are not permitted in the small 12th-century walled town itself, but that makes it particularly pleasurable to wander around, taking in the magnificent architecture; the polychrome Gothic doorway of the Romanesque Santa María de los Reyes is one of the finest in the world. Dine at Amelibia or – if you can score an invitation – La Casa Cosme Palacio (see page 126).

Laguardia

Alfaro

At the easternmost point of Rioja, Alfaro is an attractive town of substantial size, though sparsely populated. It is the last navigable point on the River Ebro (*el faro* means 'the lighthouse'), and while tourism is not a primary concern here, there are plenty of fine monuments, including the 16th-century Colegiata de San Miguel Arcángel church. Alfaro is also the hometown of Álvaro Palacios, one of Spain's most accomplished winemakers, whose extended family's empire stretches across northern Spain. Eat at Morro Tango, Palacios's favourite restaurant, and visit the excellent wine shop Ninfeo del Vino (see age 163). See also The best bodegas for tours and tastings, page 110.

The ultramodern fantasy of Hotel Viura sits almost in the lap of the ancient church of San Andrés, Villabuena de Álava

Wine routes

Although Rioja is a big region, it's relatively contained. The most famous bodegas and towns are all within an easy day's drive, and two or three days in the area will be ample for even the most dedicated Rioja lover. Over the next few pages are itineraries for day-trippers, weekenders and those driving down from Bilbao or San Sebastián.

The c19th bridge of San Juan de Ortega (known as the Puente de Piedra) in Logroño, a landmark on the pilgrim route

Despite its prominence in the wine world, Rioja is still something of a hidden gem as far as tourism is concerned, although more and more visitors are discovering the magic of the area and its wines.

Roughly 160km (100 miles) from both Bilbao and San Sebastián, Rioja is easily accessible by car and train from either; although a day trip is possible, it makes more sense to stay overnight and give yourself as much time in the region as possible.

Travelling to Rioja

Getting to Rioja is easy, although the region has no international airport. The nearest airport is Bilbao, from where you can make your way to either Haro or Logroño (1.5 hours by car; 2.5 hours by train). There are a few international flights to San Sebastián, too, which is about the same distance from Rioja.

The wine routes detailed below are in Rioja, but the journey is often as much a part of the experience as the destination, and including Bilbao and San Sebastián in your itinerary is highly recommended – both are vibrant cities with very different personalities. While Bilbao has the rugged edge of the industrial city it once was, now combined with stunning modern architecture, San Sebastián is a coastal resort with one of the most dynamic gastronomic scenes in Europe.

Even travelling between the two is beautiful: along the coastal road linking Bilbao to San Sebastián is a stretch where the waves, quite literally, meet the mountains. Stop in the charming fishing village of Getaria for turbot grilled over an open fire at Elkano, or stay overnight and get the most out of one of Rioja's deepest wine lists at Kaia Kaipe.

Day trip: Haro

There are a few railway stations that are revered in the wine world. The tin shed at Coonawarra in South Australia is one, the beautiful tiled station at Pinhão in the Douro another. But perhaps the most famous of all is the station at Haro, because this is where the story of modern Rioja began (see History, page 14). Clustered around this unassuming terminus, within an easy walk of each other, are some of the most eminent names in Rioja: CVNE, Bodegas Bilbaínas Viña Pomal, Gómez Cruzado, La Rioja Alta, Muga, Bodegas Roda and López de Heredia, with its striking new Zaha Hadid-designed tasting room. Most of these are open for visits, and many have wine bars and shops if you don't have time for a full visit (see Best bodegas for tours and tastings, page 110). Make sure you also factor in a long lunch at one of the best restaurants in Rioja, the Michelin-starred Nublo (see Where to eat and drink in Haro, page 155).

Two nights: Logroño and Haro

With a thriving restaurant scene and excellent hotels, Logroño is bigger and more vibrant than Haro, but it has fewer wineries. There's Javier Arizcuren's urban winery in the centre, nearby Bodegas Vinícola Real (vinicolareal.com) and a particularly good tour at the 130-year-old Bodegas Franco Españolas (francoespanolas.com/en/winery). In the evening, follow the town's famed tapas route (see The best tapas in Logroño, page 149).

Haro is the historical centre of Rioja, a busy, working town with a venerable history. You should take a few hours to wander its streets, taking in the Renaissance church of Santo Tomás and the wonderful Plaza de la Paz, with its colonnaded bars and tall, glass-balconied houses.

Weekend road trip: Haro and the villages

Haro – Briones – Labastida – San Vicente de la Sonsierra – Elciego – Laguardia – Logroño – San Millán de la Cogolla – Santo Domingo de la Calzada

As well as iconic bodegas, this route takes in famous stops on the Santiago de Compostela pilgrim trail and historic medieval villages. You can base yourself at a number of places along the route – at Hotel Marqués de Riscal in Elciego, for example; in the pared-back 16th-century luxury of Santa María Briones; or at the very fine, and

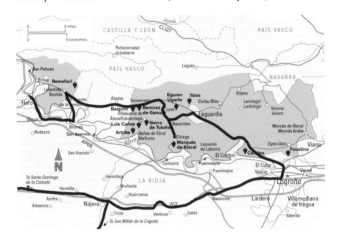

very old, **Parador de Santo Domingo de la Calzada**. From Briones you can visit the medieval walled town of **Labastida**, notable for being the birthplace of Manuel Quintano, the late-18th-century Bordeaux-loving priest credited with introducing barrel ageing to Rioja. From there, it's a short drive to **San Vicente de la Sonsierra**, built around a 10th-century hilltop fortress. Just out of town, near the Romanesque **Ermita de Santa María de la Piscina**, you can walk up to the 1,000-year-old stone wine presses of Zabala, carved into the rock – testimony to Rioja's centuries-old wine tradition. Before heading back to Haro, stop off at the **monasteries of Suso and Yuso** at **San Millán de la Cogolla**, where monks are said to have set down the first words in Castilian and Basque.

A long weekend: Logroño to Alfaro
Logroño – Calahorra – Quel – Autol – Aldeanueva de Ebro – Alfaro

Heading east out of Logroño, your first stop on this route is Marqués de Murrieta (marquesdemurrieta.com). Its top cuvée, Castillo Ygay, comes from a single 40-hectare (100-acre) vineyard, La Plana, planted in 1950. In 2011, the 160-year-old bodega – Castillo Ygay – was taken down to its foundations and entirely rebuilt to the original design, but with new tasting rooms, a museum, and a 70,000-bottle library of Murrieta vintages going back to 1852.

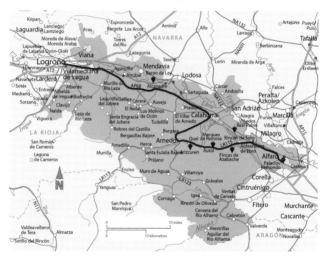

From Murrieta it's about half an hour to Calahorra, the de facto capital of Rioja Oriental. There are some fine monuments here, but Calahorra is worth a visit simply to experience an unadorned, working northern Spanish town.

At Quel, Arizcuren Bodega's four-hour tour of its Sierra de Yerga vineyards and 15th-century castle features a series of tastings, including one on top of a hill, and a traditional lunch. For details on Quel's **Bread and Cheese Festival**, see Visiting Rioja, page 84.

Autol is home to another of Rioja Oriental's high-quality cooperatives. Marqués de Reinosa (marquesdereinosa.com) has 1,200 hectares (2,965 acres) of vineyards, the majority old bush vines. It is well worth a visit to get an idea of the dramatic difference between the terroirs of this part of Rioja and the lands to the west.

In **Aldeanueva de Ebro**, Fincas de Azabache is a major co-op with 2,600 hectares (6,425 acres) of vines. Visiting this major producer allows you to see another side of Rioja production from a state-of-the-art winery. Aldeanueva also has a small wine museum in the 17th-century Ermita de Nuestra Señora del Portal.

Alfaro's most famous son is Álvaro Palacios, but his Palacios Remondo winery is not open for visits (see Towns and villages, see page 92). There are, however, several interesting family-run bodegas around Alfaro that give an insight into this less-visited corner of Rioja. At **Viñedos Real Rubio** (realrubio.es), which has been in the same family since its founding in the early 20th century, it's possible to see some very old bush vines.

Rioja and the north coast: Bilbao – San Sebastián - Rioja

By combining your trip to Rioja with stays in Bilbao and San Sebastián, you will be able to enjoy the beautiful north coast, see the Guggenheim and eat some of the best seafood in Spain. A longer stay in the region will also give you time to explore the villages of Rioja Oriental, particularly some of the cooperatives – a great way to understand the vital role they play in wine production here.

Bilbao has changed hugely over the past 25 years. Once a rugged industrial port town, it's now one of Europe's most vibrant cities, home to the Frank Gehry-designed Guggenheim Museum (see Marqués de Riscal, pages 108–109), as well as a number of excellent restaurants. An hour by car to the east is Getaria, where Pedro Arregui's Michelin-starred Elkano serves a fire-grilled turbot so famed that it has been copied by celebrity chefs from Manchester to

Essex. Then it's a short hop to San Sebastián – but avoid August, when temperatures soar and the city is choked with traffic. (San Sebastián is a much-loved holiday spot for the Spanish.) There, you have the choice of some internationally renowned restaurants, as well as seafront bars and restaurants elbowing each other for space. Check out Mugaritz (mugaritz.com), the grand-daddy of tasting-menu blowouts; **Txepetxa** (bartxepetxa.es) for superb pintxos; the hip Casa 887 (grupo887.com); modern pintxo pioneers **Sukaldean** (sukaldeanaitorsantamaria.com); **Antonio Bar** in the Parte Vieja; and **Zapiain** (zapiain.eus) to try the delicious home-brewed cider.

From San Sebastián, it's a pleasant drive south to Rioja. True sybarites (with deep pockets) might consider looking in on **Asador Etxebarri** (asadoretxebarri.com) in the village of Atxondo, surrounded by woodland, roughly equidistant between Bilbao and San Sebastián and about two hours by car from Logroño. Etxebarri, where chef Victor Arguinzoniz pioneered the art of cooking over flame, has been on 'world's best restaurant' lists since the 2010s. Be warned, though: the booking system is opaque at best.

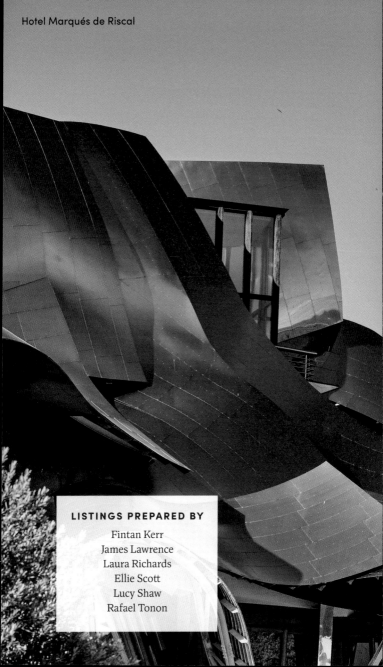

Hotel Marqués de Riscal

LISTINGS PREPARED BY

Fintan Kerr
James Lawrence
Laura Richards
Ellie Scott
Lucy Shaw
Rafael Tonon

The Guide

Contents

The best bodegas for tours
and tastings   110

The best wine hotels   122

Fine dining in Rioja   134

The best bodega
restaurants   140

The best tapas in
Logroño   149

Where to eat and
drink in Haro   155

The best wine shops
in Rioja   162

Wines to look out for   166

CLUB OENOLOGIQUE

The Guide is produced in
partnership with Club
Oenologique, the premium lifestyle
publication connecting people to
the joys of the world through the
lens of wine and spirits

Scan for the latest
on Rioja, and
expert guides to
food and drink
regions around
the world

The best bodegas for tours and tastings

Tastings in some of the world's most awesome cellars, bike or Segway tours, lunches with your own personal chef or picnics in a hot-air balloon floating above the vineyards – Rioja offers some unforgettable bodega visits. Here are some of the best.

The best way to discover a wine region is to visit it. This is especially true of Rioja now, because the region has undergone incredible change since the turn of the 21st century: new, exciting producers are popping up in tastings around the world, and classic winemakers are finally having their moment in the sun. The picturesque winemaking villages and towns are becoming destinations for wine lovers keen to learn more about the region.

'Exploring the wineries and tasting the wines with the people who made them is an opportunity that shouldn't be missed'

In turn, Rioja's tourism infrastructure is catching up with demand. Pioneers such as Marqués de Riscal and Edmond de Rothschild (owner of the deluxe Palacio de Samaniego) have given the area a much-needed boost with what today's wine drinker desires: understated and relaxed luxury. There is also an expansive line-up of activities on offer – from cycling tours to bespoke wine lunches with your own personal chef. You can even float above the Tempranillo vines in a hot-air balloon, with picnic included. Diversity is Rioja's greatest strength: journey back in time in the ancient cellars of CVNE – the winery boasts an unrivalled collection of vintages from the 19th century – or experience the striking modernity of Roda's chic tasting room.

Walking the vineyards, exploring the wineries and tasting the wines alongside the people who made them is an opportunity that should not be missed. And while learning about the relationship between variety, soil and climate is already rewarding, doing so in the shadow of the majestic Sierra de Cantabria, in the heartland of exceptional Spanish wine, is an experience that is hard to beat. The list that follows includes some of the very best visits and tastings from across Rioja.

Abeica

Camino San Roque 8,
26339 Ábalos

Abeica is a small project currently managed by the fourth generation of the Medina family in Ábalos, a small village near San Vicente de la Sonsierra. Rural and charming, Abeica might be off the beaten path, but it is worth visiting for its variety of tours – from a straightforward tasting, to a guided trek, and even the chance to see the traditional stone *lagares* of San Vicente de la Sonsierra, a remnant of the village's ancient past. At 500m (1,640ft) above sea level, the views across the region are stunning.

bodegasabeica.com/
enoturismo-2

Arizcuren

Calle Santa Isabel 10,
26002 Logroño

Javier Arizcuren is one of the most exciting figures in Rioja today and a leading light of the region's new wave. His focus is very much on the vineyards of Rioja Oriental. You can visit his urban winery in Logroño, or embark on an adventure to the Yerga mountain range to walk the vineyards. Either way, there's a lot of good wine and an introduction to the new style of Rioja that's rising up.

arizcurenvinos.com/en/wine-tourism

Baigorri

Vitoria–Logroño km 53,
01307 Samaniego

Baigorri, much loved by followers of terroir-driven Riojan wine, is one of the region's most welcoming destinations. The winery is a striking example of modern architecture, with concrete and glass blending seamlessly into the landscape of Rioja Alavesa. Its on-site restaurant serves such regional delights as oxtail pie or fish soup with cuttlefish, monkfish and clams, but those on a guided tour can also indulge in a selection of pintxos prepared by the chef. Tours are offered in Spanish, English, French and German, giving the winery a multinational feel.

bodegasbaigorri.com

Baigorri

Eguren Ugarte

Crta. A-124 km 61,
01309 Laguardia

In terms of a luxury wine experience, there is more than a whiff of Napa Valley in the air at Eguren Ugarte – its splendid hotel, complete with gourmet restaurant, offers spectacular views of the Sierra de Cantabria. In fine weather, though, most visitors make a beeline for the outdoor terrace, sipping old vintages and devouring fresh pintxos overlooking the vines. A horse-drawn carriage ride through the vineyards is popular with couples, while younger drinkers tend to plump for the Segway tour. Bespoke and private visits can also be arranged for those who want that extra touch of luxury.

egurenugarte.com

Finca Valpiedra

El Montecillo s/n,
26360 Fuenmayor

Owned by the Martínez Bujanda family, Finca Valpiedra, on a bend of the River Ebro, is a rising star on the Spanish wine scene. Its state-of-the-art bodega produces distinctive Riojas in a modern style: fruit and oak are harmoniously integrated in cuvées that range from a barrel-aged white to an exquisite red reserva. One of its most popular excursions

Bodegas Campillo

Bodegas Campillo

Logroño Errepidea 3,
01300 Laguardia

Established in 1990, Campillo quickly built a reputation as an excellent source of modern-style Rioja, rich in fruit. It lies at the heart of the Rioja Alavesa subregion, in the pretty village of Laguardia – a worthy attraction in itself. After a leisurely lunch, head to the winery to experience its innovative aroma masterclass, designed to enhance your olfactory skills, before tasting a range of Campillo vintages. The fee includes a guided tour of the cellar and an apéritif, which is usually something chilled and/ or sparkling.

bodegascampillo.com/en

involves a tour of the vineyards and surrounding countryside in a 4x4 jeep, followed by a riverside picnic lunch.

familiamartinezbujanda.com/en/finca-valpiedra

La Rioja Alta

Avda. de Vizcaya 8, 26200 Haro

Situated next to the historic train station at Haro, La Rioja Alta was founded in 1890 by the coming together of five winemaking families, and it remains one of the most renowned wineries in the region. Its long-aged and long-lived wines – particularly the Ardanza reserva and the 904 and 890 grandes reservas – are essential reference points for understanding what traditional Rioja is all about. The winery is also home to one of the most impressive barrel rooms in the wine world; 30,000 barrels – all produced in-house – are used for the ageing of the wines.

riojalta.com/en/wine-tourism/rioja-alta

Marqués de Murrieta

N-232a km 402, 26006 Logroño

Marqués de Murrieta is one of Rioja's most historic and respected bodegas. Its founder, Luciano de Murrieta, was the first person to export Spanish wine in barrels, and its magnificent Castillo Ygay Gran Reserva is Rioja's seminal single-vineyard wine. You can sample this iconic super-cuvée, and many more, in the bodega's spectacular tasting room or glamorous wine bar. There's also a plethora of tours available, including a cellar visit and gourmet tasting menu, paired with either four or seven wines.

marquesdemurrieta.com/en

Marqués de Murrieta

La Rioja Alta

Marqués de Riscal
Calle Torrea 1, 01340 Elciego

There is no more recognisable winery in Rioja, or perhaps the whole of Spain, than the titanium, Frank Gehry-designed Marqués de Riscal. That one of the oldest wineries in Rioja, established in 1858, has one of the most modern and exciting buildings tells you everything you need to know about the producer: tradition combined with evolution. A visit here can take many forms, including staying at the property itself. There is a Michelin-starred restaurant on-site, a spa and, of course, fantastic wines available to taste in a variety of formats.

marquesderiscal.com/en

Muga
Avda. Vizcaya 2, 26200 Haro

Few wineries combine tradition with modern progression quite as seamlessly as Muga, located in Haro's historic Barrio de la Estación ('Station Quarter'). Some of the most highly rated wines in the entire region hail from the 25,000-sq-m (270,000-sq-ft) winery, which focuses on full-bodied, long-lived styles of wine, aged in both French and American oak. The wines can be experienced at a tasting, as part of a longer visit or, quite uniquely, during a hot-air balloon ride above the vineyards of Rioja.

bodegasmuga.com/en/wine-tourism/activities

CVNE
Barrio de la Estación S/N – 26200 Haro – La Rioja

Founded in 1879, CVNE (Compañía Vinícola del Norte de España) is one of the handful of noble old bodegas that cluster around the station at Haro, in Rioja Alta. Its fame rests on a stable of wines – including the famously long-lived Imperial, Viña Real, Monopole and Contino – produced under the CVNE umbrella. Unusually for Rioja, the company owns almost all its vineyards. As the 5th generation of the family, Victor Urrutia's ambition is to own all his vines, and he continues to buy land. CVNE's warehouses and cellars are extensive, and tours are tailored to individual needs, taking visitors past giant open-topped oak fermenters and through the impressive barrel room. Many of the bodega's elegant and mellow wines can be enjoyed with a traditional Rioja style menu. If you're staying longer in Rioja, you can also visit the fine old estate at Viñedos Contino in Laserna and Viña Real in Laguardia. Be sure to book ahead for all visits, though.

cvne.com

Content supported by CVNE

Ramón Bilbao

Avda. Santo Domingo 34,
26200 Haro

Ramón Bilbao was founded in 1924, and a new bodega was built in a classical style in 1972. Standard tours here last an hour and take visitors through the winery from barrel to bottle, finishing with a small tasting of three wines. Foodies can choose between a wine-and-chocolate pairing option and the opportunity to imbibe velvety reservas alongside a selection of cheeses from Sierra de Los Cameros. There's also an enticing wine bar, restaurant (complete with *terraza*) and well-stocked shop.

bodegasramonbilbao.com/es-es

Remírez de Ganuza

Calle Constitución 1,
01307 Samaniego

Founded in 1989 by an intrepid Navarran, Remírez de Ganuza is one of Rioja's finest artisan bodegas. Producing a broad range of wines, it has nonetheless resisted the temptation to expand too far. Its hospitality offering reflects that ethos: tours are limited to small groups and must be reserved in advance. There is also a shop selling a back catalogue of venerable vintages, including the much-hyped 2004.

remirezdeganuza.com

Roda

Avda. Vizcaya 5, 26200 Haro

Roda sure knows how to charm its visitors. Guests might take part in a thought-provoking tour of the cellars and vineyards, before heading to a magnificent balcony overlooking bustling Haro; or sip voluptuous reservas and enjoy *jamón serrano* and *pan con tomate* with Roda's own delicious olive oil at the on-site restaurant. More elevated options include carriage tours through the vineyards, picnics within the grounds and even vertical tastings of rare vintages.

roda.es/en

Remírez de Ganuza

Roda's barrel room

Valdemar

Camino Viejo 24, 01320 Oyón

One of the most important producers in Oyón, Álava, Valdemar is a pioneer of Rioja, and it was the first winery in the region to produce a barrel-fermented white wine, as well as a Garnacha gran reserva. The focus here is heavily on the vineyards – all 275ha (680 acres) of them. Visits are small and personal, despite the large scale of the winery, and they can be tailored according to group size and experience level. It's a fantastic way to go behind the scenes with a modern Rioja trailblazer.

valdemarfamily.com/en/visit-us

Valenciso

Ctra. Ollauri–Nájera km 0.4, 26220 Ollauri

One of the greatest wineries flying just under the radar is Bodega Valenciso, founded in 1998 in Ollauri. The name is an amalgamation of the surnames of the two owners, Luis Valentín and Carmen Enciso. This is very much in the new wave of Rioja, and the winery is a marvel to behold – no expense has been spared in its making. The focus here is on old-vine Tempranillo from some of the coolest sites in Rioja, often planted at high altitude, and the wines express all the verve and depth that are increasingly sought in this region's offerings. Tours can be booked as part of a group at set times or privately, and there is also a wine bar on the premises.

valenciso.com/en/content/8-visits

Viña Real

Ctra. Logroño-Laguardia km 4.8, 01300 Laguardia

Viña Real takes a 'progressive traditionalist' approach to winemaking, producing an eclectic mix of classic and modern styles, including some truly beguiling single-estate wines. Situated at the heart of the Rioja Alavesa subregion, the winery is flanked by the captivating Sierra de Cantabria: fertile Instagram territory. Guided tours are available, and many choose to add a leisurely lunch into the bargain. The bodega also offers picnics among the vines and vineyard tours, complete with wine, cheese and chorizo.

cvne.com/bodegas/vina-real

The bottle cellar at Viña Real

Ysios

Ysios

Camino de La Hoya s/n,
01300 Laguardia

In Rioja Alta, framed by the Cantabrian mountains in the background, stand the vineyards and winery of Ysios. Designed by Santiago Calatrava, the winery building, with its iconic undulating roof, was inspired by the mountains, and architecture fans should find as much to enjoy here as lovers of powerful reds. A range of visits is available – from a tour rooted in architectural appreciation, to a guided trip through the vineyards in a specially adapted 'wine car'. Should you be visiting on a whim and without a reservation, Ysios is also home to a fantastic wine bar.

bodegasysios.com/en/visit-the-winery

Where winemakers dine

I'd choose **Morro Tango**, in my hometown of Alfaro. Cristóbal Castillo's cuisine is modern but with a great respect for our gastronomic traditions. He cooks classic dishes with refinement and a contemporary touch, and his flavours have a clean, bright purity to them. From stews, peppers and artichokes, to fish and meats, Castillo combines ingredients through a local lens, and he has an incredible knowledge of fine dining, thanks to the decade he spent at Echaurren. Don't leave without trying the croquettes, which pay tribute to Doña Marisa Sánchez. The phrase *morro tango* refers to those who like to eat well – it's something our grandparents used to say to us.

Álvaro Palacios,
Palacios Remondo

The best wine hotels

From the strikingly mismatched wood, glass and concrete boxes of the Viura hotel, designed by Joseba and Xabier Aramburu, to Frank Gehry's titanium wonder at Marqués de Riscal, Rioja seems to cast a spell over architects. And nowadays, most wineries have hospitality in mind...

Finca de los Arandinos

The north of Spain is not just for oenophiles; it's also a playground for design aficionados interested in locating structures that have sprung from the minds of some of the most renowned architects working today.

Wine tourism kicked off in Rioja with the much-hyped 2006 opening of Marqués de Riscal's City of Wine, which also features a Caudalie vinotherapy spa and a quartet of restaurants. Riscal's bold bet on tourism set a trend. While some of the top names open their doors only to members of the wine trade, most of the major bodegas in Rioja now offer at the very least tours and tastings, while many have expanded to include contemporary restaurants, boutique hotels and a host of visitor experiences – from horse-drawn carriage rides to Segway tours through the vineyards.

Several new hotels have opened in Rioja Alta and Alavesa, with Haro, Laguardia and Logroño at the epicentre of the movement. Just as Rioja is celebrated for its long-aged grandes reservas and polished *vinos de autor*, it also caters to those who want to sleep somewhere sleek and modern.

Over the following few pages, you'll find ten of the best places to stay in Rioja – from the Rothschild-owned Palacio de Samaniego to the ultra-exclusive, invitation-only La Casa Cosme Palacio.

Hotel Viura

Marqués de Riscal

Torrea 1, 01340 Elciego

Marqués de Riscal changed the tourism game in Rioja in 2006 when it opened its glinting wonder of a wine hotel designed by Canadian-American architect Frank Gehry, whose wave-like design was inspired by the folds of a flamenco skirt. Behind the cascading sheets of purple, gold and silver titanium lie 61 guest rooms and suites with panoramic views over the medieval town of Elciego. Given the contours of the building, each of the rooms is unique in shape and character, with high ceilings and huge windows looking out through a web of steel supports.

For those looking to wind down, there's a Caudalie vinotherapy spa – complete with indoor pool and fitness centre – where guests can indulge in grape scrubs and moisturizing

Marqués de Riscal

wine wraps. Keen to cater to every whim, the hotel is home to four restaurants: an alfresco *asador* (grill); a casual wine bar and bistro; restaurant 1860 Tradición, which celebrates the best of Basque cuisine; and the Michelin-starred Restaurante Marqués de Riscal, headed up by chef Francis Paniego, where you'll find many back vintages of Riscal on pour alongside 200 other top names. Top tip: don't miss sundowners at the rooftop lounge bar.

marquesderiscal.com/en/
marques-de-riscal-hotel

Hotel Viura

Herrería 19A,
01307 Villabuena de Álava

Standing next to a 17th-century sandstone church in the tiny medieval village of Villabuena de Álava, Hotel Viura is set against a backdrop of the Cantabrian Mountains. The cubed rooms of this four-star luxury boutique hotel designed by architects Joseba and Xabier Aramburu pile nonchalantly on top of each other. It may look like an office block fallen from the sky, but somehow it works.

Named after Rioja's flagship white grape, Viura offers 35 spacious, minimalist rooms with huge beds. Everything is cool – from the red Nespresso machines to the chic bath

Hotel Viura

Where winemakers dine

There are many restaurants that I could recommend, but I'd have to choose **Amelibia** in Laguardia. The cuisine is quite traditional, but the main courses have an avant-garde touch. My favourite is roast pork on a bed of sour apples, which is a perfect mix of the tenderness of the pig and its crunchy crackling with the freshness of the apples. The wine list brings together the labels of the area, but if you want to enjoy special wines, ask Alejandra for her secret notebook, and she'll let you in on a few treats.

**Barbara Palacios,
Bodega Barbara Palacios**

products. Its rooftop lounge bar has an outdoor cinema and offers 360-degree views of the surrounding area. The cellar, meanwhile, includes a sizeable offering of barrel-fermented Viuras and a number of old vintages of Rioja big guns, including CVNE, Marqués de Riscal, López de Heredia and Roda. In the restaurant, gold barrels hover from the ceiling, and dishes such as cod croquettes, crab ravioli and green pea and black truffle pintxos are Basque-inspired with a modern twist.

hotelviura.com/en

La Casa Cosme Palacio

Calle San Lázaro 1,
01300 Laguardia

Taking bespoke stays up a notch is the by-invitation-only La Casa Cosme Palacio, inside 19th-century winery Bodegas Cosme Palacio. Designed to feel like a home from home, the guesthouse is modern but retains the building's original features. Service is next level: clients have their own round-the-clock chef and butler, and they are free to roam the 4,000-sq-m (43,000-sq-ft) grounds. While there, guests can ride an e-bike through Tempranillo vineyards, indulge in beauty treatments, work up an appetite at the fitness centre, take a dip in the outdoor pool, and enjoy a private tour of the winery, followed by a wine-pairing dinner whipped up by an in-house chef.

The guesthouse majors in privacy and comfort. Designed by architect Gregorio Marañón, landscape gardener María Medina and interior designer Cristina Arechabala, the stylish, Scandi-inspired space creates a sense of calm in harmony with its surroundings. It boasts nine bedrooms and four suites decked out in elegant, neutral tones, with open brickwork and wooden furniture adding to the homely vibe.

lacasacosmepalacio.com/en/home

La Casa Cosme Palacio

Finca de los Arandinos

Ctra. LR–137 km 4.6,
26375 Entrena

Rioja does modernism well, and the hotel at Bodega Finca de los Arandinos is no exception. Perched on a hilltop in Entrena, near Rioja's capital of Logroño, the glass-fronted Cubist structure looks like something you might see on the Cornish coast. The winery and airy rooms above it were designed by architect-turned-winemaker Javier Arizcuren, embracing an industrial-chic aesthetic through a mix of concrete and blond wood. Across the lobby you'll find ten smaller rooms with modular furniture and daring decor by Spanish fashion designer David Delfín. All rooms offer tranquil views of the vine-studded landscape.

The tasting-menu-driven in-house restaurant Tierra (meaning 'earth' or 'land') champions hyper-local ingredients; they are given a contemporary twist in dishes such as turbot in a green Thai sauce, and steak tartare with foie gras shavings. The cherry on the cake is the hotel spa, where guests can detox in the sauna and steam room, lather up in a Turkish bath, top up tans on sun loungers made from barrel staves, then cool off in the pool.

fincadelosarandinos.com

Eguren Ugarte

Crta. A–124 km 61,
01309 Laguardia

Wake up to views of vines as far as the eye can see at Eguren Ugarte. Spread across spacious grounds, the hotel – unmistakable from the outside thanks to a turret jutting out of its roof like a medieval watchtower – includes a golf course, miniature lake and restaurant. The interiors offer the perfect combination of traditional flourishes – decorative tiles and wall murals – and modern twists, from plush beds and spacious showers to a complimentary mini-bar stocked with beer, soft drinks and water.

Eguren Ugarte

Where winemakers dine

I was going to talk about **Nublo** and its sister restaurant **Los Caños**, but I'm sure others will say plenty about them. Also in Haro is the less showy but equally magnificent **Terete**. Leaf through a Spanish restaurant guide from the 1980s, and Terete is there among Spain's finest. Go for the lamb, of course, but also the *merluza frita*, which is excellent. In fact, it's all good. Terete is a family business, like every good Spanish restaurant. Jaime, who runs the floor, is charming, knowledgeable and the keeper of many old vintages of great Haro wines. Go during the week, though, to avoid the crowds.

**Victor Urrutia,
CVNE**

The hotel pays homage to winery founder Vitorino Eguren, and each room is furnished with his 200-page memoir, which guests can fan through on a private balcony overlooking the Toloña Mountains. On-site Restaurante Martín Cendoya (see page 145) specializes in vine shoot-grilled lamb chops and other Rioja classics – all paired with wines from the bodega's huge back catalogue.

egurenugarte.com/bodega-con-hotel-rioja-alavesa

Finca La Emperatriz
Ctra. de Santo Domingo, Haro, 26241 Baños de Rioja

At Bodegas Finca La Emperatriz, overnight visitors enjoy the privacy of villa life, with a trio of properties available to rent a stone's throw from the vines. This was once the property of Eugenia de Montijo, the last empress of France and wife of Napoleon III. The former farm workers' cottages have been given a modern makeover while preserving their original features, and they provide the perfect setting for rest, relaxation and a digital detox, complete with stunning views over the Cantabrian and Demanda mountains.

Two of the villas have two bedrooms, making them ideal for family stays; the other is a

Palacio de Samaniego

solo act aimed at couples. Each has its own private terrace with vineyard views. During a stay, which includes a complimentary continental breakfast, guests can go on tours of the vineyards, winery and cellars, and swot up on the Finca La Emperatriz range of wines during a tutored tasting, followed by sundowners at the estate's alfresco wine bar.

hermanoshernaiz.com/en/
villas-eng

Palacio de Samaniego

Constitución Kalea 12,
01307 Samaniego

This stylish residence has the revered Rothschild name behind it. In 2004, the famous wine family partnered with Ribera-based Bodegas Vega Sicilia on a Rioja project called Macán, which includes a cutting-edge winery in Samaniego. Three years ago, the family added a boutique hotel into the mix, inside an 18th-century palace decorated with artworks from Baroness Ariane de Rothschild's private collection. Its nine brightly coloured rooms, set around the building's original stone staircase, are named after different grape varieties.

Small details are a big deal here – from the complimentary bottles of wine on arrival, to the pillow menu. The hotel's suite comes with a fitted kitchen, while on-site restaurant Tierra y Vino fuses the best of French and Basque cuisines. Chef Bruno Coelho shines a light on local produce, and when he's not manning the stoves you'll find him tending to his veg patch in the kitchen garden. Don't leave without enjoying an e-bike ride through the vineyards, followed by a snooze in a hammock on the terrace.

palaciodesamaniego.com/en

Casa Robla

Calle Hornos 11, 01309 Elvillar

Housed in a former winery in Rioja Alavesa, pared-back Casa Robla, originally built in the 18th century, was sensitively restored in 2022 by an architectural studio based in Madrid. You get a sense of this story through thoughtful and occasionally striking design details – from the subtle transformation of old glass demijohns into decorative lighting features, to the dramatic positioning of an ancient wine press in the communal breakfast room. Not only was upcycling key to Casa Robla's restoration, but the eight-bedroom property is also powered by renewable energy.

Casa Robla is owned by winemaker Cristina Blanco – who steers the ship at nearby Bodegas Blanco Pérez de Azpillaga – and her husband Eduardo Zabala and is operated as a B&B. With their clean lines, natural materials and muted tones, the en-suite rooms are almost monastic in their simplicity – and intentionally so. While there's no restaurant on-site, it's only a 10-minute drive to Laguardia, where plenty of tapas bars await.

casarobla.com/en

Santa María Briones

Calle Concepción 37,
26330 Briones

Set in a sympathetically renovated 16th-century manor house, boutique hotel Santa María Briones dials up the charm. Its 16 rooms are all large, soothing spaces with original stone walls, painted wooden beams, L'Occitane bath products and vineyard views. Ask for the Berones suite, which is split over two levels and offers direct access to the manor's main balcony, or for room 33, which has the best views. The hotel overlooks the Ebro river and village of Briones, with its ancient defensive wall, which the courtyard is built around.

Guests can take their pick from two restaurants: Allegar for fine dining, and El Calado for traditional-leaning cuisine. At the former, Rioja native Juan Cuesta puts a modern spin on local classics – try the pickled partridge ravioli, for example. Look out for his crowd-pleasing honeyed crispy

Casa Robla

Santa Maria Briones

lamb on sweet potato cream and panko breadcrumbs. At the atmospheric El Calado, expect rustic yet refined dishes like grilled Iberian pork and Basque cheesecake. Work off all the indulgence at the gym, or keep the good times going at the library-style wine bar.

santamariabriones.com/en

Hospedería del Vino

Bodegas Puelles, Camino de los Molinos, 26339 Ábalos

This former hotel – one of Rioja's first wine hotels – has been transformed into a guesthouse (sleeping 12) that can be rented in its entirety for family stays or weekends with friends. Boasting six double rooms with private bathrooms and balconies, in addition to a spa with heated pool, sauna, Jacuzzi and sun terrace, it is ideal for those seeking a self-contained stay.

Open March to October, the pet-friendly guesthouse has a fully equipped kitchen and a dining room that can accommodate 20. The rooms are bright and cosy, with a pared-back aesthetic alongside modern touches such as a smart TV and mini-bar. Best of all, they have floor-to-ceiling windows that swing open to offer views across the vineyards. All guests receive a guided tour of the winery and grounds, which includes a tasting of five wines, including the bodega's 100% Viura, crianza and reserva. Look out for the 17th-century mill during the vineyard tour.

bodegaspuelles.com/reservar-alojamiento-en-bodega-de-la-rioja

In the kitchen at the Michelin-starred Nublo in Haro

Fine dining in Rioja

Riojans like to eat well, and adventurously, and they have the choice of numerous top-class restaurants with excellent wine lists – from the young superstars of Ikaro to the traditional (and locally renowned) Casa Toni.

El Portal de Echaurren

The region is blessed with abundant natural resources – from the Atlantic Ocean to the north for the freshest fish, to the cool, green pastures where some of Spain's most famous beef is raised. On Rioja's doorstep is the Basque Country – Spain's culinary hotspot – which means that there is always cheffing talent passing through. Perhaps most important, though, is the cultural element. Riojans like to eat, and they like to eat well. With such an embarrassment of riches on their doorstep, it's easy to see why.

Traditionally, food in Rioja has been fairly rustic, focused on slow-cooked meats, stews and other hearty fare that pairs not only with the local wines but also with the continental climate. However, as the wine industry has grown, so too has the demand for a more sophisticated approach to food. Taking a leaf out of the Basque Country's book, the region's fine-dining scene has flourished.

'Riojans like to eat, and they like to eat well. With such an embarrassment of riches, it's easy to see why'

Today, it is possible to eat well almost anywhere in Rioja. Even the smaller wine towns and villages have restaurants that take great care and attention when creating their menus, which are focused on fresh, seasonal produce. Better yet, the scene appears to have grown at the same rate as modern wine appreciation, leading to a number of established restaurants with excellent wine lists to complement the delicious food.

While not exhaustive, the following list covers some of the very best fine dining in the Rioja region. *¡Buen provecho!*

Ajonegro

Ikaro

Avda. Portugal 3 bajo,
26001 Logroño

Iñaki Murua and Carolina
Sánchez managed to claim their
first Michelin star within only
18 months of opening Ikaro, a
fusion restaurant in the heart
of Logroño. Murua is from
Laguardia, in Rioja Alavesa, and
Sánchez comes from Ecuador,
and while their food is focused
on local ingredients, it is also
coloured by the experience and
heritage of these two young
superstars. Their creativity
is abundantly clear across a
number of menus that vary in
length and include suggested
wine pairings.

restauranteikaro.com

Ajonegro

Calle Hermanos Moroy 1,
bajo 9, 26001 Logroño

Ajonegro is a charming
restaurant with a Spanish-
Mexican fusion concept – no
doubt the influence of its
two chefs: Mariana Sánchez
from Cuernavaca, Mexico,
and Gonzalo Baquedano
from Logroño, who worked
together at the famous ABaC in
Barcelona. Two evening tasting
menus are offered with a choice
of wine pairings, and there is an
à la carte option that combines
the likes of mole and tacos with
fresh oysters and steak tartare.

restauranteajonegro.com

Aitor Esnal

Calle Sagasta, 13, 26001
Logroño España

Aitor Esnal is the third Logroño restaurant of Basque husband and wife team Aitor Esnal and Beatriz Martinez, the chef and sommelier behind Marinée and Wine Fandango (see page 153). An excellent wine list, three tasting menus (one in homage to Esnal's mother), additional Asian influences and a cool open-plan dining room and kitchen make this one of the most exciting new restaurants (it opened in 2020) in Rioja's vibrant foodie landscape.

aitoresnal.com/restaurante

Venta Moncalvillo

Ctra. Medrano 6,
26373 Daroca de Rioja

If there's a must-visit restaurant for wine lovers, it's the two-Michelin-starred Venta Moncalvillo in Daroca de Rioja, southwest of Logroño. Ignacio Echapresto manages the menu, with many of the ingredients sourced from the restaurant gardens, while brother Carlos handles the front of house and arguably the greatest wine cellar in Rioja. With more than 600 offerings, including a great deal of old, rare Rioja, this is the ideal restaurant to visit for a splurge.

ventamoncalvillo.com/web

Head chef Ignacio Echapresto
at Venta Moncalvillo

Where winemakers dine

My recommendation is **El Portal de Echaurren**. Because it is located in Ezcaray, the town where I was born, it has been the scene of many happy moments in my life, including during my childhood. The restaurant has been open for 125 years, and it was the first in Rioja to obtain both one and two Michelin stars, which helped to put Riojan gastronomy on the map. Echaurren is driven by a respect for tradition and a quest for innovation, which is demonstrated in both its dishes and its magnificent wine list. The croquettes are still faithful to the restaurant's original recipe and are not to be missed. The ambience is warm and cosy, like a home from home, and the wine list boasts great names from Rioja, the rest of Spain and around the world. The best time to visit is in the spring.

Mayte Calvo de la Banda Salgado, Bodegas Bilbaínas

Casa Toni

Calle Zumalacárregui 27, 26338 San Vicente de la Sonsierra

In the village of San Vicente de la Sonsierra, with its winding, medieval streets and walled facades, is Casa Toni. The atmosphere here is warm and inviting, and the food is traditional. Roasted pigs' feet, sausages, potatoes and chorizo are staples on the menu, though there are also many contemporary dishes. Much of the wine list hails from nearby producers, making for a perfect introduction to the village as a whole.

casatoni.es

La Cocina de Ramón

Calle Portales 30, 26001 Logroño

Less than 100 metres from a local market, La Cocina de Ramón puts fresh produce front and centre. Ramón Piñeiro – along with his hard-working team – works the region's native vegetables into a great many of the recipes. His delicious, hearty food is ideal after a long day of travelling and drinking wine. The restaurant, just a few streets away from Logroño's cathedral, is small and popular – a combination that suggests advanced bookings are a smart idea.

lacocinaderamon.es

La Vieja Bodega

Avenida de La Rioja 17, 26230 Casalarreina

As recommended by Esperanza Tomás of Roda (see page 158), La Vieja Bodega is charmingly old-fashioned. In a former 17th-century bodega, the high-ceilinged, stone-walled and raftered dining room is cosy and intimate; the food is traditional/contemporary influence – there's sushi on the menu, alongside such local delicacies as rabo estofado (oxtail stew) and a variety of classic lamb dishes. The 600-strong wine list is famously reasonable – Viña Tondonia Reserva 2013 is €85 for example (it will cost you more in a shop); by-the-glass starts at €3.

laviejabodega.es/en/home

Restaurante Allegar

Calle Concepción 37, 26330 Briones

Briones isn't just a famous winemaking village; it's also considered one of Spain's most beautiful medieval villages. Restaurante Allegar, in Hotel Santa Maria Briones, is its finest restaurant. The name translates as 'to scrape one's plate clean', which sets the tone for an indulgent dining experience. Allegar offers a modified take on Riojan cooking with a few twists, all the while allowing the essence of local ingredients to deliciously shine through.

santamariabriones.com/gastronomia

El Portal de Echaurren

Calle Padre José García 19, 26280 Ezcaray

The first restaurant in Rioja to receive a Michelin star (it now has two), El Portal de Echaurren is the jewel in the crown of Ezcaray, along with sister site Echaurren Tradición. Chef Francis Paniego is the driving force behind both venues, and he reaches his creative zenith at El Portal. The menu changes frequently, though the famous *croquetas* alone are worth the trip. For many, this is the peak of dining in Rioja. Place yourself in the capable hands of the restaurant staff, and let them guide you on both the food and the incredible wine list.

echaurren.com/en/el-portal-de-echaurren-restaurant

El Portal de Echaurren

The best bodega restaurants

Like Spain's wines, the country's cuisine seems to encapsulate perfectly a national obsession with respecting the past while also trying to reinvent it.

The dining room at Castillo Ygay

Spain is home to some of the most creative and compelling minds in modern gastronomy – from Ferran Adrià, who changed the global culinary landscape with his groundbreaking restaurant El Bulli, to the Roca brothers at El Celler de Can Roca in Girona, via Basque legends such as Juan Mari Arzak and Eneko Atxa. While France may be the cradle of classic cookery, Spain is where people come to be excited and challenged by what they eat, and it remains among the most innovative and interesting countries for its food and wine.

Those planning a trip to Rioja won't be disappointed on either front, since the region boasts an array of restaurants running the gamut from experimental to traditional. There are fine-dining venues, as well as spots specialising in local, seasonal dishes that are often cooked over fire or roasted over vine shoots for added flavour. Winery restaurants are a relatively recent phenomenon in Rioja. The trend was spearheaded by Marqués de Riscal, which in 2006 opened its ambitious hotel and City of Wine venue; it has no fewer than four on-site restaurants catering to different moods and occasions. Other bodegas soon wanted in on the action, and the Rioja region is now home to a thriving winery-restaurant scene, offering visitors the opportunity to dine among the vines in scenic settings.

True to Spanish form, the restaurants are wildly different in nature. Some, like Martín Cendoya at Eguren Ugarte, proudly fly the flag for traditional Riojan cuisine and keep things simple, serving the likes of lamb cutlets roasted over vine shoots. Others, such as Restaurante Marqués de Riscal, push the boundaries of Basque cuisine via daring dishes that marry unexpected ingredients like squid and toffee. Whether you're in the mood for a fuss-free lunch in a pretty vineyard spot, or are seeking to challenge your taste buds with outlandish flavour combinations, the following guide to the best winery restaurants in Rioja has you covered. Just remember to bring your appetite…

Restaurante Marqués de Riscal

Calle Torrea 1, 01340 Elciego

Marqués de Riscal sets the bar high when it comes to top-end gastronomy in Rioja. Housed within Frank Gehry's striking space – a symphony of silver and purple titanium folds – are a quartet of restaurants: the alfresco Asador Torrea grill; the casual Vinoteca & Bistro; 1860 Tradición, which celebrates the best of Basque and Riojan cuisine; and the Michelin-starred Restaurante Marqués de Riscal. The latter, overseen by Rioja-born chef Francis Paniego, is the jewel in the Riscal crown, thanks to its contemporary twist on traditional flavours and cutting-edge dishes that pair unexpected ingredients – such as marinated duck with praline and mushrooms, and squid tartare with toffee.

Restaurante Marqués de Riscal

Paniego, who also runs the two-Michelin-starred El Portal de Echaurren in Ezcaray, works largely with local ingredients – such as the white grape Viura in his scarlet prawn-head sauce – but takes inspiration from world cuisines. Both his seven- and ten-course tasting menus seek to encapsulate Rioja on a plate. As for the wine list, Spain is certainly championed, but it also includes top estates from around the globe.

marquesderiscal.com/en/restaurants

Restaurante Baigorri

Ctra. Vitoria-Logroño km 53, 01307 Samaniego

Built by Basque architect Iñaki Aspiazu, Bodegas Baigorri is a sleek, glass-fronted Bond-villain lair of a building, boasting jaw-dropping views across Rioja Alavesa. In something of a plot twist, the eponymous restaurant within this architectural wonder of glass and steel is located six floors underground, where arched windows allow diners to gawp at the beauty of the landscape on one side, and sleeping barrels on the other.

The restaurant's menu changes every month and includes four Baigorri wines, while the premium menu features a quintet of the bodega's top pours, such as its silky, violet-scented Garnacha. Highlighting local seasonal ingredients, chef Maite Barruti's dishes – which might include mango and cucumber gazpacho, prawn tartare with avocado and courgette, and oxtail pie with Parmentier potatoes and sautéed mushrooms – are fresh and indulgent in equal measure. Menus can be adapted for groups if organized in advance, and all visits include a winery tour.

bodegasbaigorri.com/en/reserva-restaurante

Restaurante Baigorri

Tierra

Tierra

Finca de los Arandinos, Ctra.
LR-137 km 4.6, 26375 Entrena

Taking the Spanish word for 'earth' or 'land' as its name, Tierra, housed within the glass-fronted, Javier Arizcuren-designed Finca de los Arandinos, champions hyper-local ingredients. The restaurant boasts views over the vines at the foot of the Moncalvillo Mountains, and it offers up a series of tasting menus, with one option serving as a selection of the restaurant's best dishes, including white asparagus with paprika emulsion; cod flakes and black olives; and beef sirloin with truffled Parmentier potato, beef jus and a cured egg yolk.

Spanish ingredients are given an avant-garde twist via daring dishes such as turbot in a green Thai curry sauce, and steak tartare with foie gras shavings and a crispy corn tortilla. And be sure to check out the caramelized French toast with vanilla ice cream.

fincadelosarandinos.com/
restaurante-tierra

La Casa Cosme Palacio

Calle San Lázaro 1,
01300 Laguardia

Epitomizing the current trend for bespoke experiences, Cosme Palacio's guesthouse – La Casa Cosme Palacio, in Rioja Alavesa – can be booked only by invitation. Those lucky enough to stay there will be put in the safe hands of the in-house chef, who will cater to their every culinary whim, serving bountiful alfresco barbecues in the summer months, alongside meals in a cosy, stone-clad dining room, softened by a pale-blue-and-white colour scheme and pared-back Scandi aesthetic. With an open fire roaring in the middle of the room, the dining space is where Cosme Palacio's grapes used to arrive fresh from the vines on large carts.

Many of the ingredients that go into the dishes are sourced from the bodega's kitchen garden. Each of the dishes on the tasting menu – which currently includes spider crab with Basque beef emulsion; creamy rice with baby squid; and grilled fig with fig-leaf ice cream – is paired with a different wine from the range; lobster stew is served with golden glasses of Cosme Palacio Blanco 1894. During harvest, you can lunch among the vines and taste the first must from a traditional press.

lacasacosmepalacio.com/en/la-casa

Restaurante Martín Cendoya

Crta. A-124 km 61, 01309 Laguardia

Shining a light on traditional Riojan fare is this restaurant, part of the family-run Eguren Ugarte winery in the town of Paganos, near Laguardia, Rioja Alavesa. At Martín Cendoya, chefs cook over wood in open fires, offering a pared-back menu that highlights quality ingredients. Kick off proceedings with some gooey

La Casa Cosme Palacio

Idiazabal cheese croquettes, then move on to lamb chops roasted over vine shoots served with a simple lettuce, tomato and onion salad and Riojan string beans.

The restaurant champions Basque specialities from *pochas* (white beans) to homemade pastries – and it weaves *morcilla* (blood sausage) into tortillas oozing with *salsa riojana*, a local red pepper and tomato sauce. More modern dishes include grilled octopus with white truffle and paprika, and clams with salsa verde. Even lighter fare – like the wonderfully decadent fried eggs with caviar – delivers on indulgence.

egurenugarte.com/ experiencias-enoturismo-rioja-alavesa/#restaurante

Restaurante Gonzalo de Berceo

Ctra. Villar de Torre s/n, 26310 Badarán

Combining fine dining with literature, the subterranean restaurant at Bodegas David Moreno pays tribute to 12th–century Spanish poet Gonzalo de Berceo, who hailed from Rioja and is celebrated as an early pioneer of Castilian verse. The cavernous space boasts a vaulted ceiling and arched windows that overlook the barrel cellars, giving diners an immersive experience.

Taking a local approach, the restaurant specializes in meats roasted over vine shoots from its Badarán vineyard to enrich their flavour. Paired with a simple garden salad, the lamb cutlets cooked this way are not to be missed. Keeping things traditional, the restaurant also goes big on roasted suckling lamb – a perfect pairing for red Rioja – and Riojan-style potatoes served with chorizo. A guided tour of the winery is offered to all diners.

tienda.davidmoreno.es/ restaurante

Restaurante Martín Cendoya

Castillo Ygay

Castillo Ygay
Private Kitchen

Bodega Marqués de Murrieta,
N-232a km 402, 26006 Logroño

When the King of Spain officially inaugurates a restaurant, you know it's going to be a big deal – and the private kitchen at Marqués de Murrieta doesn't disappoint. Housed within the rebuilt Castillo Ygay, whose decade-long renovation necessitated some 6,000 tonnes of stone, the restaurant serves gourmet tasting menus that celebrate clashes as much as complementary flavours.

Including a winery tour, the dining experience has been created with the wines in mind; Marqués de Murrieta's 70,000-strong cellar boasts bottles dating back to 1852. The ten-course menu features four wines, including Marqués de Murrieta Limited Edition 2014 and Capellanía 2019, served alongside dishes made with Spanish ingredients but tipping their hat to French techniques – such as foie gras terrine with an orange coulis, and roasted sea bass in a citric velouté. The 14-course extravaganza, meanwhile, pulls out the big guns, with wines on pour including Champagne Gosset Grande Réserve Brut and Castillo Ygay 2012. Its dishes range from scarlet prawn tartare with aromatic coral, to stuffed artichoke gratin with olive and soy, finishing with an almond soufflé with honey ice cream.

reservasenoturismo.
marquesdemurrieta.com/
reserva/visitas

Restaurante Vivanco

N-232 km 442, 26330 Briones

Vivanco has been spreading the good word of Rioja since 2004, when it opened its Museum of Wine Culture. But man cannot survive on knowledge alone, so the bodega is also home to a circular restaurant with impressive panoramic views over the vineyards and the Cantabrian Mountains. Championing local produce cooked in a wood-fired oven and on grills fuelled by vine twigs, rootstocks and oak from old wine casks, chef Maribel Frades riffs on traditional Riojan dishes with contemporary flourishes.

With close attention paid to texture and presentation, there are several tasting menus on offer – from the six-course Riojan Heritage Menu, to Harmonies, which includes sommelier-selected wine pairings. Standout offerings include *morcilla* croquettes with roasted red peppers, cod loin in a garlic and red chilli emulsion, and spiced French brioche toast with grilled pear and Tempranillo ice cream. There's also a tapas bar on the terrace for those seeking light bites overlooking the vines.

vivancoculturadevino.es/en/experiences/restaurant

Where winemakers dine

The place everyone is talking about in Rioja is **Nublo** in Haro (see page 157). Miguel Caño, the chef and owner of this 16th-century building, worked on the research team at Mugaritz. His team at Nublo is full of ambition and offers a unique fine-dining experience. It supports the young winemaking community and local producers, with a commitment to wines from villages and growers that will one day be Rioja's next iconic producers. The place is amazing, and dining there gives you a sense of the history of the space.

Telmo Rodríguez, Remelluri

The best tapas in Logroño

Calle del Laurel, in Rioja's provincial capital Logroño, has become an unmissable tapas destination. Walking down this street, as well as those adjacent to it, visitors get a real feel for the energy of the city.

Logroño, the tapas trail

On any given summer evening, elderly couples might enjoy a glass of wine, while young families introduce the newest member to *croquetas de jamón iberico*. Young locals spend the first half of their night out eating, drinking and catching up around Calle del Laurel before moving on to a club or party.

In modern Spain, eating tapas is part of the normal way of life. The verb *tapar* means 'to cover'; the tradition of small bites is said to have originated in the south of Spain, where small pieces of bread, sometimes with a topping, would be given out as a cover for filled glasses to stop flies from landing in the wine.

Where better than in the heart of Rioja to indulge in a dining tradition that has its roots in wine drinking? In Logroño, as in other parts of the country, the custom of moving from bar to bar has led to establishments differentiating themselves through their specialities. From juicy mushrooms dripping with herbs and butter at Bar Soriano, to tender, delicious *tortilla de patatas* at El Canalla, there are very good reasons to keep moving and tasting.

This is also a fantastic place to enjoy local wines by the glass, with some of Rioja's most interesting wine bars springing up in the area. You can't go far wrong in Logroño, but here are some of the bars you really shouldn't miss.

Bar Jubera
Calle del Laurel 18

Founded in 1980, Bar Jubera has become one of the most popular destinations on Calle del Laurel for its *patatas bravas* – a dish that is meant to be spicy but rarely is. Thankfully, Bar Jubera's version delivers the heat. While *bravas* sauce is normally made using a healthy dose of smoked paprika, at Jubera they use a local chilli to give it extra oomph. This classic dish done well makes for an excellent way to line the stomach for what's to come and is the perfect start to an evening exploring the area.

Where winemakers dine

My favourite Rioja restaurant is **Asador Tres Tinas** in the town of Viana. I like it because it only has a few tables, and the atmosphere is very personal. Whenever I go, I let them choose what I eat. I tend to begin with seasonal starters such as piquillo peppers and a *cogollos de tudela* salad combining anchovies with cod, scrambled eggs and mushrooms. The main courses are prepared on the grill right in front of you, so you get a front-row seat to the action. They serve tasty fish like grilled sea bream, sea bass and hake. The grilled meats, such as beef sirloin steaks and lamb chops, are also delicious. Don't leave without ordering a dessert – the praline Mozart cake with mandarin sorbet is particularly good.

Simón Arina, Baigorri

El Canalla
Calle Albornoz 1

Another institution of the Logroño tapas scene is El Canalla, whose speciality is an exemplary *tortilla de patatas*. Made with baked Álava potatoes, caramelized onions and eggs, it is so soft that it's almost unnecessary to use a knife when divvying it up. Given the small, cosy nature of El Canalla, expect a warm and atmospheric welcome to accompany the comfort food – or perch at one of its outdoor tables on warmer evenings.

Roots
Calle Marqués de Vallejo 14

Everybody on the Logroño tapas trail eventually seems to find themselves at Roots, regardless of their original plans. It's one of the newest wine bars in the city, and it's certainly the most varied, serving an unusually large number of wines by the glass, both Riojan and international. As a result, it's not unusual to see local winemakers catching up and comparing their wines with those made by their peers all over the world. Don't sleep on the excellent tapas range, though, with local meats and cheeses on the menu, plus anchovies and *gildas* skewers with hake.

rootsrioja.com

Roots

Wine Fandango
Calle Vara de Rey 5

Wine Fandango is a collaborative project between four individuals: two from the world of gastronomy and two from the world of wine. Located in the former legendary Grand Hotel, overlooking the Paseo del Espolón park, Wine Fandango is more of a restaurant than a bar, with a complete gastronomic offering that includes small bites on the go with a glass of wine – or three. Keep an eye out for their creamy *croquetas* stuffed with *jamón ibérico*. Come on Saturday afternoons for a weekly party with live DJ music.

winefandango.com

Bueno Bueno
Calle San Juan 33

A little off the beaten path, Bueno Bueno is a traditional tapas bar that's already laying a claim to fame for its *lacón*, a salty, smoky meat taken from the shoulders or front legs of Iberian pigs. Sliced into thin pieces and served with a little olive oil, paprika and crusty bread, *lacón* works wonderfully with a glass of juicy, sappy Tempranillo.

Pata Negra
Calle del Laurel 24

This is one of the most popular stops on Calle del Laurel, and with good reason: there's a huge range of freshly made pintxos (like tapas, but served on a skewer on top of a piece of bread), as well as one of the city's best selections of wines by the glass. This is a wonderful all-rounder that is constantly busy and takes no reservations. Arrive early, grab a stool and soak up the atmosphere.

Pata Negra

Torrecilla
Calle del Laurel 15

Another icon of this famed *calle*, Torrecilla has a varied selection of tapas and pintxos, which changes on a regular basis, based on available local ingredients. Heading in for a glass of wine or a beer, you'll be overwhelmed by a choice of bites that typically cost around €3, such as foie toast, mini hamburgers, Moorish skewers, omelettes and strips of sirloin. The main difficulty here is trying to stick to a single choice.

Cafeteria Tizona
Calle Ciriaco Garrido 14

Some bars are famous for their history or reputation, while others are renowned for certain tapas or pintxos. Cafeteria Tizona can claim both as the 2022 winner of one of the country's most prestigious culinary awards: Spain's Best Tortilla. With a soft, gooey centre, this is not an item to eat on the go; instead, savour it with some soft, slightly crunchy bread before moving on into the night.

instagram.com/bartizona

Bar Soriano
Travesía del Laurel 2

Affectionately known as the *campeones de champiñones* ('mushroom champions'), Bar Soriano specializes in one thing only: a towering mushroom-based pintxo gently scented with garlic and local herbs. Needless to say, there's plenty of excellent wine to wash it down with. This is one of the most famous stops for tapas in the Laurel neighbourhood – a must-visit.

'Bar Soriano specializes in one thing only: a towering mushroom-based pintxo gently scented with garlic and local herbs'

Where to eat and drink in Haro

A surge in wine tourism is under way in Haro. Standing on the banks of the River Ebro, this charming town has a rich history dating back to the 10th century. It also features one of the world's largest concentrations of century-old wine cellars.

Labyrinthine stone streets in Haro's picturesque old quarter lead to new restaurants and upscale hotels that tie in with the artisanal vision of emerging winemakers. Fuelled by a rising interest from international travellers, a rich culinary landscape has developed, ranging from haute cuisine establishments to beloved traditional eateries.

'The hospitality scene has transformed local life,' says chef Miguel Caño, who decided to return to his hometown to open the restaurant Nublo. 'The season is not just limited to summer and the harvest; we have visitors all year round.' Caño has also been responsible for modernizing his family's establishment, Los Caños, to reflect the evolution of Haro's culinary scene, which is fundamentally rooted in Riojan tradition.

Across the town, locals and visitors alike gather at unassuming tapas bars and cosy family-run restaurants renowned for exceptional cuisine. The following is a selection of the best places to eat and drink in Haro.

Nublo

Los Caños

Nublo
Plaza San Martín 5

Housed in a 16th-century mansion, this Michelin-starred restaurant (the first and only, so far, in Haro) is a testament to the culinary excellence of chef Miguel Caño, formerly of Mugaritz. With just 30 seats, the intimate setting, crafted by modernist designer Santos Bregaña, seamlessly blends contemporary forms with the original historic building, ensuring an unforgettable dining experience. The tasting menu, based on seasonal and locally sourced ingredients, evolves daily to showcase the freshest offerings from surrounding farmers. Cooked in a wood-fired oven and grill fuelled by a blend of vine shoots, oak and haya wood, the food delivers an irresistible symphony of smoke and flames, elevating the region's wines to new heights.

nublorestaurant.com/en

Los Caños
Plaza San Martín 5

Adjacent to Nublo is the restaurant that ignited the Caño family's enduring legacy on the Haro fine-dining scene. Established in 1930, Los Caños quickly became a culinary haven in the city, first under the guidance of chef Miguel Caño's grandmother and later his father. Upon his return to his hometown, Caño embarked on a renovation project, revitalizing

the space and infusing the menu with a contemporary twist while preserving its traditional essence. The focus is on local and seasonal ingredients used across comfort dishes such as *bonito ventresca* tuna with prawns and eggs, *pollo al ajillo* (garlic chicken), and crispy hake tempura served with homemade mayonnaise. The wine list showcases a unique selection of lesser-known Rioja labels alongside the renowned ones, while the warm service makes you feel like you're dining in a cherished family home.

loscaños.es

El Rincón del Noble
Calle Martínez Lacuesta 11

Rioja's culinary tradition is alive and well at El Rincón, where the local cuisine has been practised for generations. For decades, the restaurant has attracted guests keen to savour its renowned *chuletillas de cordero* – moist lamb chops grilled on embers. This is *tradición riojana* at its finest – from *ciervo estofado* (braised venison), to duck confit with caramelized apple, and *manitas de cerdo en salsa riojana* (pigs' trotters in Rioja sauce). The stone walls, with their murals of the vineyards, exude an old-world charm and warmth.

elrincondelnoble.net

Terete

Calle Lucrecia Arana 17

One of Haro's most iconic eateries, this restaurant draws food enthusiasts seeking tender roasted leg of lamb cooked to perfection in a wood-fired oven, and a flavourful rendition of *menestra de verduras*. Here, the *menestra* is a type of stew featuring a medley of seasonal vegetables (such as asparagus, peas, artichokes, olives, carrots, potatoes, green beans and onions), with delicate slices of Serrano ham to enhance the flavour. Housing bottles cherry-picked from the region, the cellar, in the lower part of the restaurant, dates back to the end of the 19th century, when it was established alongside the lamb roastery. Terete has persevered through many decades, and today the family's fifth and sixth generations are committed to upholding their culinary heritage with unwavering passion and expertise.

terete.es

El Trujal del Abuelo

Calle Bodegas 1, 26210 Cihuri

This hidden gem sits next to a Roman bridge crossing the River Tirón in Cihuri, west of Haro. Housed in a charming old building, the small family restaurant is quaint, but the cuisine is exceptional, showcasing the best products cooked over vine shoots on an open-air *parrilla* grill. Indulge in signature dishes like the charcoal-grilled Iberian *secreto*

Terete

Alboroque

(a cut located in the back part of the pork loin), served with peppers and artichoke hearts, or enjoy the exquisite smoked anchovies.

instagram.com/explore/ locations/1015026736/ restaurante-el-trujal-del-abuelo

Bodega Pimiento
Calle Salvador 8, 26211 Tirgo

The name of this restaurant in Tirgo, a ten-minute drive from Haro, derives from one of the region's most cherished culinary traditions, a custom dating back to the 19th century. In autumn, when red and green peppers reach their peak ripeness, Riojans gather to roast and savour them. This simple yet exquisite dish is served here with chorizo and a fried egg. The menu also includes succulent grilled *chuletillas* (lamb chops roasted on vine shoots) and Riojan-style potatoes, offering a delightful complement to the region's renowned wines. For dessert, try the stuffed puff pastries, rice pudding or cheese with quince marmalade – all provide an irresistible finale to a memorable dining experience.

instagram.com/elpimientotirgo

Alboroque
Plaza de la Paz 8, Haro

In a charming building in Haro's central square, Alboroque offers a sophisticated dining experience with meticulously crafted dishes that emphasize the quality of local ingredients. The menu, though concise, caters to all palates, with options such as caramelized *pimientos del piquillo* (bell peppers) with smoked cod, crispy *morcilla* with a cream of capers, or braised oxtail with potato cream and vegetable couscous. Of course, lamb is also a highlight and is served daily for lunch. The restaurant is open for dinner on Saturdays only.

restaurante-alboroque.makro.rest

Chamonix
Calle Santo Tomás 14–16

Located in Haro old town, Chamonix is one of the many bars that contribute to the vibrant atmosphere of the La Herradura neighbourhood, named after its horseshoe shape and encompassing the streets of San Martín and Santo Tomás. Chamonix is a beloved local institution renowned for its tapas and wines, often produced just a few kilometres – or even metres – away. The ambience evokes that of a cosy mountain retreat, complete with a glass display showcasing the day's freshly prepared snacks, such as grilled mushrooms, artichokes, and *pinchos morunos* (lamb skewers), a Moorish-inspired delicacy similar to a kebab. For heartier appetites, the homemade hamburger is an excellent choice.

chamonixharo.es

Where winemakers dine

Alameda in Fuenmayor is a traditional restaurant owned by husband-and-wife team Tomás and Esther. Based on top-quality seasonal produce, dishes are deeply rooted in northern Spanish gastronomy. It's one of those venues where you can have an honest experience of time and place. Tomás is a master with the grill and gives the likes of Asador Etxebarri a run for its money. You can enjoy both fish and meat from San Sebastián here, the latter from famous butcher Patxi Larrañaga in Lasarte, who supplies the top restaurants in Gipuzkoa. Esther is in charge of the rest of the menu. Start with the mouthwatering croquettes, and if you dine in spring, don't miss the *guisante lágrima* (teardrop peas), white asparagus and broad beans. During the summer months, be sure to order the tomato salad; in autumn, boletus; and during winter, thistle and borage.

**José Urtasun,
Remírez de Ganuza**

The best wine shops in Rioja

There's a particular thrill to finding a well-stocked wine merchant in wine country – one whose hidden corners might conceal rare old vintages and long-lost treasures. Even better is that many of these shops don't share their inventory online, meaning that the very best bottles aren't snapped up remotely by collectors. The only way to know what's in stock is to pay the shop a visit.

Rioja is a region with no single dominant retail outlet. Instead, a number of small, privately owned shops can be found in Haro and Logroño and in the surrounding villages. What follows here is a selection of the best. Of course, you can buy wine directly from the bodegas, but for a broader selection, independent advice and the chance of finding a gem, these shops are well worth a visit. (See also Great Rioja vintages, page 80.)

Ninfeo del Vino

Avda. Zaragoza 4, 26540 Alfaro

Alfaro is one of the key winemaking villages of Rioja Oriental and home to five wineries. Ninfeo del Vino is one of the best wine shops in Rioja – if not in Spain – with a selection of hand-picked wines from across Rioja Oriental, as well as an impressive range of local gastronomic products. Owner Juan Navajas is extremely knowledgeable, and the selections in store are entirely down to his decades of experience in Rioja.

ninfeodelvino.com

Viura Negra

Calle Marqués de Murrieta 5, 26001 Logroño

Vinoteca Viura Negra is perhaps best known as a gourmet store, but it also has a huge and well-chosen wine selection. The customer service here might be the very best in Rioja: knowledgeable, attentive and personable. Look out for the organized wine tastings as well – they are a highlight of any week and a great addition to a Rioja wine itinerary.

viuranegra.com

Ninfeo del Vino

Where winemakers dine

Located in an imposing 16th-century palace in Haro, **Nublo** (see page 157) is a fascinating restaurant. Miguel Caño's gastronomy has a strong identity, focusing on simplicity and incorporating only the finest ingredients. Miguel loves old-fashioned slow-cooking techniques, and his chefs work with fire in all its variations. He also champions local producers and showcases only local Rioja wines on his menu. Nublo's menu varies depending on what they have in that day, which is part of the restaurant's charm. I love that there is no set menu and that each time you go you'll have a new and different experience.

**María Vargas,
Marqués de Murrieta**

Vinoteca Larría

Vinoteca Larría
Avda. Pérez Galdós 52,
26002 Logroño

A well-established name in Logroño, Vinoteca Larría is one of the town's largest and best-stocked wine merchants. Proprietor Antonio León can often be found telling keen customers about his offerings, each of which he chooses with consideration and care. The store stocks more than 350 wines, predominantly from Rioja but also with a sizeable amount of Ribera del Duero and a small international section. Best of all is the collection of small grower-producers, including some little-known names. If you are unsure what you're looking for, put yourself in Antonio's expert hands. He also ships within Spain at very affordable rates.

larria.es

La Monumental de los Vinos

Avda. Santo Domingo 7, 26200 Haro

La Monumental de los Vinos is a Haro landmark. It carries one of the broadest selections of wine in Rioja, as well as its own-labelled (and very cheap) Tempranillo. Owner Julián Marroquín isn't shy about airing strong opinions on his range. While there is not a huge amount of older vintages on offer, he stocks very good new-wave wines, particularly those made in a lighter style through carbonic maceration. The other highlight is a great selection of cured meats – *longaniza* and chorizo, in particular. Although the shop is a little outside the centre, it is definitely worth a visit.

monumentalvinos.com

La Tienda de la Rica

Avda. Portugal 15, 26001 Logroño

La Tienda de la Rica is the brainchild of Javier and Arturo Amurrio Barroeta, Luís López González and Ignacio Uruñuela, four men who work together at Finca de la Rica in Labastida. Their vision was simple: to open a shop that celebrates the individuality of Rioja and the region's smaller projects, which sometimes struggle to be visible.

This is an outlet for a new style of Rioja; the owners are passionate about the new *vino de viñedo singular* classification that was introduced in 2017. Here, wine lovers can discover what is fresh and exciting in the region, courtesy of people who have worked here their entire lives.

latiendadelarica.com

Vinum Vita

Calle Virgen de la Vega 15, 26200 Haro

Vinum Vita is both shop and wine bar – and a very good bar at that. The focus is on an excellent selection of Rioja, and wines are available by the glass from an Enomatic machine. This makes for a perfect try-before-you-buy system. It's also possible to settle in and order a full bottle to drink on the premises.

instagram.com/vinum_vita_

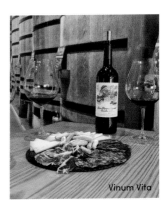

Vinum Vita

Wines to look out for

Compared with wine regions of similar history and renown, Rioja represents incredibly good value. Although prices have increased in the past few years as the region's reputation has grown, you can still find bottles with decades of life ahead of them for a fraction of the cost of a Bordeaux or Burgundy.

From the cool high-altitude vineyards in Rioja Alavesa, to the old bush vines of Rioja Oriental, the region produces a vast range of different wines and styles, many of which have yet to make it into the mainstream wine media. Value is not about low prices, but more a reflection of quality-to-price ratio and interest in a global marketplace.

Garnacha For many years, Garnacha was considered a workhorse grape – particularly in Rioja, where it was mostly used for cheaper wines and for adding heft to blends. Since the 1990s, huge swathes of Garnacha vineyards have been uprooted in favour of the more bankable Tempranillo. Recently, however, the grape has undergone something of a global renaissance – Australian Grenache, for example, is now considered as capable of complexity as Pinot Noir – and Rioja's winemakers have been quick to exploit this new-found popularity. Rioja has many old vineyards of high-quality Garnacha, particularly in Rioja Oriental. Producers such as Palacios Remondo, Bilbaínas, Miguel Merino, Contino and Arizcuren are all making Garnachas that are winning plaudits while staying at a sensible price point.

Contino Garnacha 2019

Bodegas Bilbaínas Viña Pomal Vinos Singulares de Garnacha 2017

Arizcuren Solo Garnacha Amphora 2019

Miguel Merino La Ínsula 2019

Sierra de Toloño La Dula Garnachas de Altura 2022

Vinos en Voz Baja Costumbres Tinto 2020

'The whites can be vibrant and textured – a far cry from the powerfully oaked wines of a generation ago'

White Rioja Perhaps the greatest value of all in Rioja is found in its white wines. They make up a small proportion of overall production, and they can be vibrant and textured – a far cry from the powerfully oaked wines of a generation ago. Viura, Tempranillo Blanco, Malvasia and Garnacha Blanca make the majority of the really interesting wines, both as single-variety expressions and as complex blends. Chardonnay and Sauvignon Blanc – now permitted – have not yet been enthusiastically adopted by winemakers. Look out for wines from the likes of Abel Mendoza, Finca Allende, Martinez Lacuesta, Palacios Remondo and Bodegas Pujanza.

Bodegas Ysios Blanco 2021

Manuel Quintano La Calavera 2020

Artuke Trascuevas 2021

Abel Mendoza 5V 2022

*López de Heredia Viña Tondonia
    Reserva Blanco 2012*

Finca Allende Blanco 2019

Joven wines Much is made of the longevity of Rioja, yet the quality of the grapes going into many of the joven ('young') wines is exceptional. Not all wines are meant to be aged – indeed, much of Rioja is intended to be drunk within a year or two of vintage. These fresh, sappy wines that roll charmingly over the palate can be excellent value. A prime example is Artuke's entry-level Rioja (also known as Artuke's Joven).

Artuke 2023

Sonsierra Selección Tinto 2023

Señorío de P Peciña Tinto Cosecha 2023

Ostatu Joven 2023

Remírez de Ganuza R Erre Punto Tinto
    Tempranillo 2023

Tentenublo Custero 2022

'Savoury, earthy wines with powerful acidity and firm tannins are staples of classic Rioja'

Second wines To borrow a term from Bordeaux, 'second wines' refers to wines that don't come from the best parcels of the best vineyards, or the grapes that don't make it into the top cuvée. When made by the best producers, they can be very good indeed – and excellent value. For example, La Rioja Alta produces two grandes reservas known as 890 and 904. Complex and capable of great ageing, they are deservedly pricey. The next wine in the portfolio is the excellent Viña Ardanza, a reserva whose length of ageing would allow it to be classified gran reserva. A stunning wine, at less than half the price of its more exalted siblings.

La Rioja Alta Viña Ardanza 2016
López de Heredia Bosconia 2012
Muga Reserva Muga Selección Especial 2019
Artuke Pies Negro 2020
Señorío de P Peciña Crianza 2016
Bozeto de Exopto 2022

Classically styled wines Savoury, earthy wines with powerful acidity and firm tannins are staples of classic, traditional Rioja. Much attention is focused on the new wave of producers, but many bodegas have been making wine in the traditional style for decades. They can be underappreciated but are very good value.

Señorío de P Peciña Gran Reserva 2014

Remelluri Reserva 2016

Bodegas Riojanas Monte Real Gran Reserva 1998

Ontañón Gran Reserva 2010

López de Heredia Viña Tondonia Gran Reserva 2001

La Rioja Alta 940 Gran Reserva 2010

Monte San Lorenzo, Rioja's highest peak, rises majestically above the vineyards

Glossary

Barrel, barrica – usually American oak but French oak becoming more popular. Under ageing regulations (see Reserva etc) capacity must be 225L

Barrio de la Estación – the district, including seven wineries, around the historically important train station in Haro

Bodega – literally meaning 'wine cellar', used to refer to a winery or bottle shop

Bush vine – untrained or untrellised vines

Carbonic maceration – fermentation technique using whole grapes and carbon dioxide to give light, fruity wines

Consejo Regulador – the regulatory body in Rioja, responsible for guaranteeing the quality of Rioja DOCa wines

Cosecha – harvest or vintage, also used to label young wines with no ageing requirement

Crianza – wine aged for a minimum of two years including at least one year in 225L oak barrels (six months in oak for white and rosé)

DO/DOCa – Denominación de Origen / Denominación de Origen Calificada: a classification system in Spain which protects the geographical origin and quality of wines

Field Blend – wine made from different grape varieties planted in the same vineyard and picked and vinified together

Finca – farmhouse, estate or country house

Gran Reserva – wine aged for a minimum of five years, including at least two in oak and three in bottle (minimum four years with six months in oak for white and rosé)

Hospedería – guesthouse

Joven – literally meaning 'young', a classification of wine that does not require any ageing. Now called 'Genérico'

Lagar, Lagares – open stone troughs traditionally used for fermentation

Matador Manifesto – a 2015 document signed by 150 wineries calling for single-vineyard and village names to be permitted on wine labels

Méthode bordelaise – using small oak barrels for maturation, as in Bordeaux

Paraje – parcel of land

Pago – a small parcel of land similar to a lieu-dit

Phylloxera – plague which devastated European vineyards in the 19th and 20th centuries, caused by a louse which feeds on vine roots

Pintxos – Basque small bites similar to tapas, from the Spanish 'pincho' (spike) for the cocktail stick they often come on

Reserva – wine aged for a minimum of three years, including at least one year in oak and two years in bottle (minimum two years with six months in oak for white and rosé)

Rioja Alavesa – smallest subregion, located between the Ebro and the Sierra Cantabria

Rioja Alta – westernmost subregion stretching from west of Haro to Logroño

Rioja Oriental – easternmost, warmest and driest subregion, formerly known as Rioja Baja

Rioja n Roll – a collective of eight producers putting an emphasis on terroir

Roble – literally meaning 'oak' on a label, this shows the wine has spent an unspecified time in oak

Terroir – the combination of specific factors including soils, aspect and microclimate which give wine a sense of place. *Terruño* in Spanish

Torco – a deep fermentation tank carved into a cave in medieval times

Vat – larger oak or stainless steel vessels, usually holding thousands of litres

Vertical tasting – a tasting of different vintages of the same wine

Viñedo – vineyard

Viñedo en... – a new designation approved in 2024 to label wines as coming from a particular vineyard, even if the wine is made and bottled elsewhere

Viñas Viejas / Cepas Viejas – old vines

Viñedo singular – a high-quality category wine from specific plots. Amongst other requirements vines must be at least 35 years old and harvested by hand

Vino de autor – 'signature wine', high-quality wine usually made in limited quantities

Vino de Cosechero – wine usually made using carbonic maceration by small-scale winegrowers

Vino de pueblo – literally, 'village wine'. Wine from one of Rioja's 144 named villages, which must be produced, bottled and aged in the village. Formerly 'vino de municipio'

Vino de zona – a wine from only one subzone (Rioja Alta, Rioja Alavesa or Rioja Oriental). The wine must be produced, bottled and aged in the subzone

Further reading

From Don Quixote to Lorca, the canon of Spanish writing is vast. This is a small selection of wine books and general literature that we feel should be on any hispanophile's shelf.

Wine books

The Wines of Northern Spain: From Galicia to the Pyrenees and Rioja to the Basque Country by Sarah Jane Evans, Infinite Ideas 2018
The definitive guide from Master of Wine Sarah Jane Evans.

The Wine Region of Rioja 2ⁿᵈ Edition by Ana Fabiano, Arena Books New Paltz N.Y. 2017
An in-depth look at the region, including a food section with recipes.

The New Vignerons: A New Generation of Spanish Wine Growers by Luis Gutiérrez, Planeta Gastro 2017
Profiles of 14 winemakers across Spain.

The New Spain: A Complete Guide to Contemporary Spanish Wine 2ⁿᵈ Edition by John Radford, Mitchell Beazley 2004
A tour through Spain by one of the great writers on Spanish wine.

Books on the culinary culture of Spain

A Late Dinner: Discovering the Food of Spain by Paul Richardson, Bloomsbury 2007
Richardson's captivating memoir is a journey around Spain through its cuisine.

Grape, Olive, Pig: Deep Travels Through Spain's Food Culture by Matt Goulding, Harper 2016
A look at the culinary landscape of Spain, with beautiful photography.

Tasting Spain: A Culinary Journey by H.M. van den Brink, Haus Publishing 2017
An evocative mix of historical background and personal recollections.

Books on the history of Spain

Spain by Jan Morris, Faber and Faber, first published as *The Presence of Spain* 1964, new edition 2008
A classic by one of the greatest travel writers, still as relevant 60 years after publication.

The New Spaniards by John Hooper, first published as *The Spaniards* by Viking in 1986, revised edition 2006 by Penguin Books
An in-depth look at the making of contemporary Spain from the restoration of democracy in 1977.

España: A Brief History of Spain by Giles Tremlett, Bloomsbury 2022
From the leading *hispanista*, a beautifully-produced history from prehistoric times to the present. An excellent primer.

The Basque History of The World by Mark Kurlansky, Penguin Random House 2000
Kurlansky blends human stories with history in this portrait of a little-understood people.

Literature for hispanophiles

As I Walked Out One Midsummer Morning by Laurie Lee, André Deutsch 1969
One of *the* great travel memoirs, an inspirational journey from Somerset to Andalucia.

For Whom the Bell Tolls by Ernest Hemingway, Scribner 1940
The Civil War classic. See also *The Sun Also Rises*, *Death in the Afternoon*, and *The Dangerous Summer*.

Selected Poems by Federico García Lorca, first published 1993 by Bloodaxe Books, new edition Oxford University Press 2009
Lorca had an international reputation as a dramatist and poet for plays like Blood Wedding and Yerma when he was murdered by fascists in 1936 at the age of 38.

Winter in Madrid by C.J. Sansom, 2006
A spy novel set in post-civil war Madrid; Sansom has been compared to Graham Greene by some critics.

Soldiers of Salamis by Javier Cercas, Bloomsbury 2003
A critically-acclaimed novel by one of Spain's foremost writers, set during the Spanish Civil War, mixing fact and fiction.

Tomás Nevinson by Javier Marías, Penguin 2023
The last book by the renowned writer (Marías died in 2022) is a modern spy thriller set in Northern Spain.

Index

Abeica 95, 112
acidity 62, 64, 171
ageing 18, 28, 52, 60, 61, 62, 65
Aitor Esnal 137
Ajonegro 136
Alameda 142, 161
Alboroque 160-1
Aldeanueva de Ebro 105
Alfaro 33, 96, 105, 163
Amelibia 125
Arizcuren 54, 66, 102, 104, 112
Aro 43-4
Artadi 39, 45, 50
Artuke 49, 50, 51, 60, 66
Asador Etxebarri 106
Asador Tres Tinas 151
Autol 105

Baigorri 23, 48, 66, 112, 143
Bar Jubera 151
Bar Soriano 154
Barón de Chirel 43
Barrio de la Estación, Haro 20, 52, 90
Bideona 34, 35
Bilbaínas 20, 52, 81, 90, 102

Bilbao 101, 105
blending 27, 28
Bodega Contador 44
Bodega Pimiento 160
bodega restaurants 140-8
Bodegas Campillo 113
Bodegas Franco Españolas 102
Bodegas Juan Carlos Sancha 33
Bodegas Luis Cañas 34, 35
Bodegas Riojanas 81
bodegas tours and tastings 35, 90, 110-21
Bodegas Vinícola Real 102
Bodegas y Viñedos Labastida 93
Bordeaux 17, 18, 19-20
Briones 94, 95, 130, 139, 148
Bueno Bueno 153

Cafeteria Tizona 154
Calahorra 104
carbonic maceration 59
Casa Robla 130
Casa Toni 138
Castillo Ygay Gran

Reserva Especial Blanco 37-8, 64
Castillo Ygay Private Kitchen 147
cepas viejas 33
Chamonix 161
Chardonnay 67, 73
Cirsion 40-1
classifications 26-8
climate 47, 48, 49, 51, 54
Consejo Regulador 20-1, 27, 30-1
Contino 29, 116, 167
crianza wines 27, 28, 59
CVNE 9, 20, 39-40, 52, 65, 83, 102, 111, 116-17

DOCa 21, 49, 77, 78

Ebro river 47
Eguren Ugarte 113, 127-8, 141, 145
El Canalla 152
El Portal de Echaurren 134, 138, 139
El Rincón del Noble 158
El Trujal del Abuelo 159-60
Elciego 90, 94, 124, 142

Faustino 81, 82
fermentation 18, 61, 66
festivals 86–9
field blends 34, 65
Finca de los Arandinos 127, 144
Finca La Emperatriz 128–9
Finca Valpiedra 113–14
Fincas de Azabache 105

Garagarza, Eugenio de 17–18
Garnacha 33, 50, 54, 55, 61, 65, 66, 69, 70–1, 167
Garnacha Blanca 50, 63, 66, 67, 73
geography 46–55
Gil, Jose 66
Gómez Cruzado 20, 52, 102
good value wines 166–71
Graciano 50, 61, 69, 71
gran reserva wines 27, 28, 59, 77
grape varieties 68–73
  red grapes 69–71
  white grapes 72–3
greatest Rioja wines 36–45

Haro 9, 20, 51–2, 90, 93, 101–2, 155–61, 165
Hermanos Peciña 63
history of Rioja winemaking 14–22
Hospedería del Vino 131
Hotel Marqués de Riscal 6–7, 124
Hotel Viura 98–9, 123, 124–5
hotels 122–31

Ikaro 136

joven wines 59, 169

La Batalla del Vino 86, 88
La Casa Cosme Palacio 126, 144–5
La Cocina de Ramón 138
La Cueva del Contador 44
La Monumental de los Vinos 165
La Rioja Alta 20, 21, 38, 52, 60, 62–3, 82, 102, 114, 115
La Rioja Alta Gran Reserva 890 38
La Tienda de la Rica 165
La Vieja Bodega 139, 158

Labastida 92, 93, 103
labels 76–9
lagares 15, 16
Laguardia 50, 51, 95, 96, 107, 126, 127, 144–5
Lagunilla 63
Lalomba 24–5
Lanzaga 41
Las Beatas 41
Laventura 67
Logroño 55, 90, 93, 100, 102, 136–7, 138, 147, 149–54, 163–4, 165
López de Heredia 20, 21, 37, 61, 62, 64, 81, 83, 102
Los Caños 157–8

Mágico 42–3
Malvasia 50, 63, 66, 67, 72
map of Rioja region 10–11
Marqués de Murrieta 37–8, 63, 64, 77, 104, 114, 147
Marqués de Reinosa 105
Marqués de Riscal 18, 19, 43, 48, 51, 90, 95, 111, 115, 123, 142
Martínez Bujanda 81
Matador Manifesto 27

Maturana Blanca 67, 73
Maturana Tinta 71
Mazuelo 50, 52, 61, 65, 66, 71
Mendoza, Abel 67
Merino, Miguel 52, 66
méthode bordelaise 18
Montecillo 82
Morro Tangeo 121
Muga 20, 21, 43–4, 52, 60, 66, 83, 102, 115
museums of wine 90–1, 105

Ninfeo del Vino 163
Nivarius 67
Nublo 132–3, 148, 157, 164

oak barrels 18, 28, 60, 61, 66
old vines 32–5

Palacio, Cosme 67
Palacio de Samaniego 111, 129
Palacios, Álvaro 54, 96, 105, 121
Palacios Remondo 33, 42, 54, 67, 105
Pata Negra 153
Paternina, Federico 81

phylloxera 18–19, 20, 33
Pineau, Jean 18

Quiñón de Valmira 42
Quintano, Manuel 17, 93, 103

Ramón Bilbao 118
red wines
  modern 65, 66
  traditional 61–3
Remelluri 50, 65, 93
Remírez de Ganuza 50, 118
reserva wines 27, 28, 59
Restaurante Allegar 139
Restaurante Baigorri 143
Restaurante Gonzalo de Berceo 146
Restaurante Marqués de Riscal 141, 142–3
Restaurante Martín Cendoya 145–6
Restaurante Vivanco 148
restaurants and bars 132–61
Ribera del Duero 21, 65
Rioja Alavesa 28, 34, 47, 48–51, 67, 69

maps 48–9, 103
Rioja Alta 28, 33, 47, 51–2, 67, 69
  map 52–3
Rioja DOCa 21, 49, 67, 77, 78
Rioja DOCa Control Board see Consejo Regulador
Rioja "Enthusiast" course 31
Rioja 'n' Roll 59
Rioja Oriental (formerly, Rioja Baja) 28, 33, 47, 53–5, 69, 71
  maps 54–5, 104
Roda 20, 35, 40–1, 48, 52, 66, 102, 118, 119
Roots 152

Salceda 81
San Sebastián 101, 106
San Vicente de la Sonsierra 95, 97, 103, 138
Sánchez, Carlos 67
Santa María Briones 130–1
Sauvignon Blanc 67, 73
second wines 170
Sierra de Toloño 49, 66
single-vineyard

wines 27, 28, 52
soils 49, 52, 54
styles of wine 22,
58–67

tannins 59, 60, 61,
171
tapas 87, 90, 149–54
Tempranillo 18, 19,
22, 50, 52, 54, 61,
65, 66, 69–70
Tempranillo Blanco
72, 73
Terete 128, 159
terroir 27, 65–6
Tierra 127, 144
Torrecilla 154
tours and tastings
35, 90, 110–21
Turruntés 73

Urbina 63

Valdemar 119
Valenisco 119–20
Venta Moncalvillo
137
Verdejo 67, 73
Villabuena de Álava
34, 98–9, 124
Viña El Pisón 39
Viña Pomal 52, 90,
102
Viña Real 40, 120
Viña Real Gran
Reserva Especial
39–40

Viña Tondonia
Reserva Blanco 37
viñas viejas 33
viñedo en...
designation 27, 28
viñedo singular 27,
28, 33, 78
Viñedos de
Aldeanueva 67
Viñedos Real Rubio
105
Viñedos y Bodegas
Sierra Cantabria
42–3
vino de cosechero 15
vino de peublo
(formerly, vino de
municipio) 27, 28
vino de zona 27, 28
Vinos en Voz Baja 54
Vinoteca Larría 164
vintages 21, 80–3
Vinum Vita 165
visiting Rioja
festivals 86–9
restaurants and
bars 132–61
seasons 85–6
top things to do
90–1
tours and tastings
35, 90, 110–21
travelling to Rioja
101
wine hotels 122–31
wine routes 51,
52–3, 100–7

winemaking towns
and villages 91,
92–9
Viura 50, 52, 63, 64,
66, 67, 72
Viura Negra 163
Vivanco 22, 90–1,
148

white wines 168
modern 66–7
traditional 63–5
Wine Fandango 153
wine routes 51, 52–3,
100–7
wine shops 162–5
winemaking towns
and villages 91,
92–9

Ysios 48, 90, 91, 95,
121

Acknowledgements

The publishers have made every effort to trace the copyright holders of the text and images reproduced in this book. If, however, you believe that any work has been incorrectly credited or used without permission, please contact us immediately and we will endeavour to rectify the situation.

1br Shutterstock/Alberto Loyo, 14 Consejo Regulador, 16 Consejo Regulador, 19 Shutterstock/David Herraez Calzada, 21 Shuttersstock/Arnieby, 26 Consejo Regulador, 36 James Sturcke, 46 Consejo Regulador, 50 Consejo Regulador, 60b Consejo Regulador, 64b Consejo Regulador, 67 Consejo Regulador, 68 James Sturcke / Alamy Stock Photo, 76 Shutterstock/Ralf Liebhold, 79tl Shutterstock/CapturePB, 79cr Shutterstock/Ralf Liebhold, 80 Consejo Regulador, 82 Consejo Regulador, 83 Consejo Regulador, 84 Shutterstock/Pecold, 86 Shutterstock/lakisha beecham, 89 Andy Arthur / Alamy Stock Photo, 91 Consejo Regulador, 106, Consejo Regulador, 107 Consejo Regulador, 108-109 Consejo Regulador, 149 Ivo de Rooij / Alamy Stock Photo, 150bl Shutterstock/Tatiana Bralnina, 150br Shutterstock/Kyle8Ken, 151bl Shutterstock/barmalini, 155 Shutterstock/Colleen Ashley, 166 Consejo Regulador, 172-173 Shutterstock/Alberto Loyo.
Images of businesses are copyright of those businesses.

Also available in
The Smart Traveller's Wine Guide series

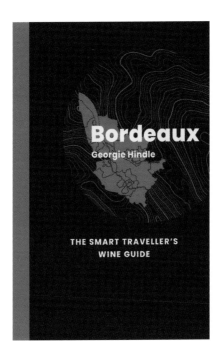

and coming soon...
Napa | Tuscany | Rhône | Switzerland

Other books from Académie du Vin Library we think you'll enjoy

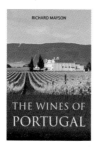

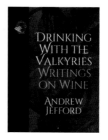

www.academieduvinlibrary.com